Status of Groundwater Quality in the Upper Santa Ana Watershed, November 2006–March 2007: California GAMA Priority Basin Project

By Robert Kent and Kenneth Belitz

A product of the California Groundwater Ambient Monitoring and Assessment (GAMA) Program

Prepared in cooperation with the California State Water Resources Control Board

Scientific Investigations Report 2012–5052

U.S. Department of the Interior
U.S. Geological Survey

U.S. Department of the Interior
KEN SALAZAR, Secretary

U.S. Geological Survey
Marcia K. McNutt, Director

U.S. Geological Survey, Reston, Virginia: 2012

For more information on the USGS—the Federal source for science about the Earth, its natural and living resources, natural hazards, and the environment, visit http://www.usgs.gov or call 1–888–ASK–USGS.

For an overview of USGS information products, including maps, imagery, and publications, visit http://www.usgs.gov/pubprod

To order this and other USGS information products, visit http://store.usgs.gov

Suggested citation:
Kent, Robert, and Belitz, Kenneth, 2012, Status of groundwater quality in the Upper Santa Ana Watershed, November 2006–March 2007—California GAMA Priority Basin Project: U.S. Geological Survey Scientific Investigations Report 2012–5052, 88 p.

Contents

Contents—Continued

Figures

Figures—Continued

Tables

Conversion Factors, Datums, and Abbreviations and Acronyms

Inch/Foot/Mile to SI

Multiply	By	To obtain
Length		
inch (in.)	2.54	centimeter (cm)
inch (in.)	25.4	millimeter (mm)
foot (ft)	0.3048	meter (m)
mile (mi)	1.609	kilometer (km)
Area		
square foot (ft^2)	0.09290	square meter (m^2)
square mile (mi^2)	2.590	square kilometer (km^2)
Radioactivity		
picocurie per liter (pCi/L)	0.037	becquerel per liter (Bq/L)

Temperature in degrees Celsius (°C) may be converted to degrees Fahrenheit (°F) as follows:

$$°F=(1.8×°C)+32.$$

Temperature in degrees Fahrenheit (°F) may be converted to degrees Celsius (°C) as follows:

$$°C=(°F–32)/1.8.$$

Specific conductance is given in microsiemens per centimeter at 25 degrees Celsius (µS/cm at 25°C).

Concentrations of chemical constituents in water are given either in milligrams per liter (mg/L) or micrograms per liter (µg/L). One milligram per liter is equivalent to 1 part per million (ppm); 1 microgram per liter is equivalent to 1 part per billion (ppb); 1 nanogram per liter (ng/L) is equivalent to 1 part per trillion (ppt); 1 per mil is equivalent to 1 part per thousand.

Datums

Vertical coordinate information is referenced to the North American Vertical Datum of 1988 (NAVD 88).

Horizontal coordinate information is referenced to the North American Datum of 1983 (NAD 83).

Conversion Factors, Datums, and Abbreviations and Acronyms—Continued

Abbreviations and Acronyms

AB	Assembly Bill (through the California State Assembly)
AL-US	U.S. Environmental Protection Agency action level
GAMA	Groundwater Ambient Monitoring and Assessment Program
HAL-US	U.S. Environmental Protection Agency lifetime health advisory level
HBSL	health-based screening level
LRL	laboratory reporting level
LSD	land-surface datum
LT-MDL	long-term method detection level
MCL-CA	California Department of Public Health maximum contaminant level
MCL-US	U.S. Environmental Protection Agency maximum contaminant level
MDL	method detection limit
NL-CA	California Department of Public Health notification level
PSW	public-supply wells
RC	relative-concentration
RPD	relative percent difference
RSD5-US	U.S. Environmental Protection Agency risk-specific dose at a risk factor of 10^{-5}
SMCL-CA	California Department of Public Health secondary maximum contaminant level
SMCL-US	U.S. Environmental Protection Agency secondary maximum contaminant level
TEAP	terminal electron-acceptor processes
TT-US	treatment technique levels
USAW	Upper Santa Ana Watershed (study unit)
USAWB	identifier prefix for grid wells in the Bunker Hill/Cajon/Rialto-Colton study area
USAWC	identifier prefix for grid wells in the Cucamonga/Chino study area
USAWE	identifier prefix for grid wells in the Elsinore study area
USAWR	identifier prefix for grid wells in the Riverside-Arlington/Temescal study area
USAWU	identifier prefix for additional wells sampled for understanding specific groundwater-quality issues
USAWY	identifier prefix for grid wells in the Yucaipa/San Timoteo study area

Conversion Factors, Datums, and Abbreviations and Acronyms—Continued

Organizations

CDPH	California Department of Public Health (Department of Health Services prior to July 1, 2007)
CDPR	California Department of Pesticide Regulation
CDWR	California Department of Water Resources
LLNL	Lawrence Livermore National Laboratory
NWQL	National Water Quality Laboratory (USGS)
SWRCB	State Water Resources Control Board (California)
USEPA	U.S. Environmental Protection Agency
USGS	U.S. Geological Survey

Selected Chemical Names

DBCP	1,2-dibromo-3-chloropropane
MTBE	methyl *tert*-butyl ether
NDMA	*N*-nitrosodimethylamine
PCE	perchloroethene (tetrachloroethene)
1,2,3-TCP	1,2,3-trichloropropane
TCE	trichloroethene
TDS	total dissolved solids
THM	trihalomethane
VOC	volatile organic compound

Units of Measure

pmc	percent modern carbon
TU	tritium unit
$>$	greater than
$<$	less than
$\leq$	less than or equal to
%	percent

Status of Groundwater Quality in the Upper Santa Ana Watershed, November 2006–March 2007: California GAMA Priority Basin Project

By Robert Kent and Kenneth Belitz

Abstract

Groundwater quality in the approximately 1,000-square-mile (2,590-square-kilometer) Upper Santa Ana Watershed (USAW) study unit was investigated as part of the Priority Basin Project of the Groundwater Ambient Monitoring and Assessment (GAMA) Program. The study unit is located in southern California in Riverside and San Bernardino Counties. The GAMA Priority Basin Project is being conducted by the California State Water Resources Control Board in collaboration with the U.S. Geological Survey and the Lawrence Livermore National Laboratory.

The GAMA USAW study was designed to provide a spatially unbiased assessment of untreated groundwater quality within the primary aquifer systems in the study unit. The primary aquifer systems (hereinafter, primary aquifers) are defined as the perforation interval of wells listed in the California Department of Public Health (CDPH) database for the USAW study unit. The quality of groundwater in shallower or deeper water-bearing zones may differ from that in the primary aquifers; shallower groundwater may be more vulnerable to surficial contamination. The assessment is based on water-quality and ancillary data collected by the U.S. Geological Survey (USGS) from 90 wells during November 2006 through March 2007, and water-quality data from the CDPH database.

The status of the current quality of the groundwater resource was assessed based on data from samples analyzed for volatile organic compounds (VOCs), pesticides, and naturally occurring inorganic constituents, such as major ions and trace elements. The *status assessment* is intended to characterize the quality of groundwater resources within the primary aquifers of the USAW study unit, not the treated drinking water delivered to consumers by water purveyors.

Relative-concentrations (sample concentration divided by the health- or aesthetic-based benchmark concentration) were used for evaluating groundwater quality for those constituents that have Federal or California regulatory or non-regulatory benchmarks for drinking-water quality.

A relative-concentration greater than (>) 1.0 indicates a concentration above a benchmark, and a relative-concentration less than or equal to (≤) 1.0 indicates a concentration equal to or less than a benchmark. Organic and special-interest constituent relative-concentrations were classified as "high" (> 1.0), "moderate" (0.1 < relative-concentration ≤1.0), or "low" (≤0.1). Inorganic constituent relative-concentrations were classified as "high" (> 1.0), "moderate" (0.5 < relative-concentration ≤1.0), or "low" (≤0.5).

Aquifer-scale proportion was used as the primary metric in the *status assessment* for evaluating regional-scale groundwater quality. Aquifer-scale proportions are defined as the percentage of the area of the primary aquifer system with concentrations above or below specified thresholds relative to regulatory or aesthetic benchmarks. High aquifer-scale proportion is defined as the percentage of the area of the primary aquifers with a relative-concentration greater than 1.0 for a particular constituent or class of constituents; percentage is based on an areal, rather than a volumetric basis. Moderate and low aquifer-scale proportions were defined as the percentage of the primary aquifers with moderate and low relative-concentrations, respectively. Two statistical approaches—grid-based and spatially weighted—were used to evaluate aquifer-scale proportions for individual constituents and classes of constituents. Grid-based and spatially weighted estimates were comparable in the USAW study unit (within 90-percent confidence intervals).

Inorganic constituents with human-health benchmarks had relative-concentrations that were high in 32.9 percent of the primary aquifers, moderate in 29.3 percent, and low in 37.8 percent. The high aquifer-scale proportion of these inorganic constituents primarily reflected high aquifer-scale proportions of nitrate (high relative-concentration in 25.3 percent of the aquifer), although seven other inorganic constituents with human-health benchmarks also were detected at high relative-concentrations in some percentage of the aquifer: arsenic, boron, fluoride, gross alpha activity, molybdenum, uranium, and vanadium.

Perchlorate, as a constituent of special interest, was evaluated separately from other inorganic constituents, and had high relative-concentrations in 11.1 percent, moderate in 53.3 percent, and low or not detected in 35.6 percent of the primary aquifers. In contrast to the inorganic constituents, relative-concentrations of organic constituents (one or more) were high in 6.7 percent, moderate in 11.1 percent, and low or not detected in 82.2 percent of the primary aquifers.

Of the 237 organic and special-interest constituents analyzed for, 39 constituents were detected (21 VOCs, 13 pesticides, 3 pharmaceuticals, and 2 constituents of special interest). All of the detected VOCs had health-based benchmarks, and five of these— 1,1-dichloroethene, 1,2-dibromo-3-chloropropane (DBCP), tetrachloroethene (PCE), carbon tetrachloride, and trichloroethene (TCE)—were detected in at least one sample at a concentration above a benchmark (high relative-concentration). Seven of the 13 pesticides had health-based benchmarks, and none were detected above these benchmarks (no high relative-concentrations). Pharmaceuticals do not have health-based benchmarks. Thirteen organic constituents were frequently detected (detected in at least 10 percent of samples without regard to relative-concentrations): bromodichloromethane, chloroform, cis-1,2-dichloroethene, 1,1-dichloroethene, dichlorodifluoromethane (CFC-12), methyl tert-butyl ether (MTBE), PCE, TCE, trichlorofluoromethane (CFC-11), atrazine, bromacil, diuron, and simazine.

Introduction

To assess the quality of ambient groundwater in aquifers used for drinking-water supply and to establish a baseline groundwater-quality monitoring program, the California State Water Resources Control Board (SWRCB), in collaboration with the U.S. Geological Survey (USGS) and Lawrence Livermore National Laboratory (LLNL), implemented the Groundwater Ambient Monitoring and Assessment (GAMA) Program (California State Water Resources Control Board, 2012, website at http://www.waterboards.ca.gov/gama/). The statewide GAMA Program currently consists of three projects: (1) the GAMA Priority Basin Project, conducted by the USGS (U.S. Geological Survey, 2010, website at http://ca.water.usgs. gov/gama/); (2) the GAMA Domestic Well Project, conducted by the SWRCB; and (3) the GAMA Special Studies, conducted by LLNL. On a statewide basis, the GAMA Priority Basin Project focused primarily on the deep portion of the groundwater resource, and the SWRCB Domestic Well Project generally focused on the shallow aquifer systems. The primary aquifers may be at less risk of contamination than the shallow wells, such as private domestic and environmental monitoring wells, which are closer to surficial sources of contamination.

As a result, concentrations of constituents, such as volatile organic compounds (VOCs) and nitrate, in wells screened in the deep primary aquifers may be lower than concentrations of constituents in shallow wells (Kulongoski and others, 2010; Landon and others, 2010).

The SWRCB initiated the GAMA Program in 2000 in response to Legislative mandates (State of California, 1999, 2001a, Supplemental Report of the 1999 Budget Act 1999–2000 Fiscal Year). The GAMA Priority Basin Project was initiated in response to the Groundwater Quality Monitoring Act of 2001 (State of California, 2001b, Section 10780-10782.3 of the California Water Code, Assembly Bill 599) to assess and monitor the quality of groundwater in California. The GAMA Priority Basin Project is a comprehensive assessment of statewide groundwater quality, designed to help better understand and identify risks to groundwater resources and to increase the availability of information about groundwater quality to the public. For the Priority Basin Project, the USGS, in collaboration with the SWRCB, developed a monitoring plan to assess groundwater basins through direct sampling of groundwater and other statistically reliable sampling approaches (Belitz and others, 2003; California State Water Resources Control Board, 2003). Additional partners in the GAMA Priority Basin Project include the California Department of Public Health (CDPH), the California Department of Pesticide Regulation (CDPR), the California Department of Water Resources (CDWR), and local water agencies and well owners (Kulongoski and Belitz, 2004).

The range of hydrologic, geologic, and climatic conditions that exist in California should be considered in an assessment of groundwater quality. Belitz and others (2003) partitioned the State into 10 hydrogeologic provinces, each with distinctive hydrologic, geologic, and climatic characteristics (fig. 1). All of these hydrogeologic provinces include groundwater basins and subbasins designated by the CDWR (California Department of Water Resources, 2003). Groundwater basins generally consist of relatively permeable, unconsolidated deposits of alluvial or volcanic origin. Eighty percent of California's approximately 16,000 public-supply wells are located in designated groundwater basins. Groundwater basins and subbasins were prioritized for sampling on the basis of the number of public-supply wells, with secondary consideration given to municipal groundwater use, agricultural pumping, the number of historical leaking underground fuel tanks, and registered pesticide applications (Belitz, and others, 2003). The 116 priority basins and additional areas outside defined groundwater basins were grouped into 35 study units for the GAMA study. These 35 study units include approximately 95 percent of public-supply wells in California's groundwater basins. The Upper Santa Ana Watershed (USAW) study unit is located in the Transverse Ranges and selected Peninsular Ranges hydrogeologic province (fig. 1) (Belitz and others, 2003).

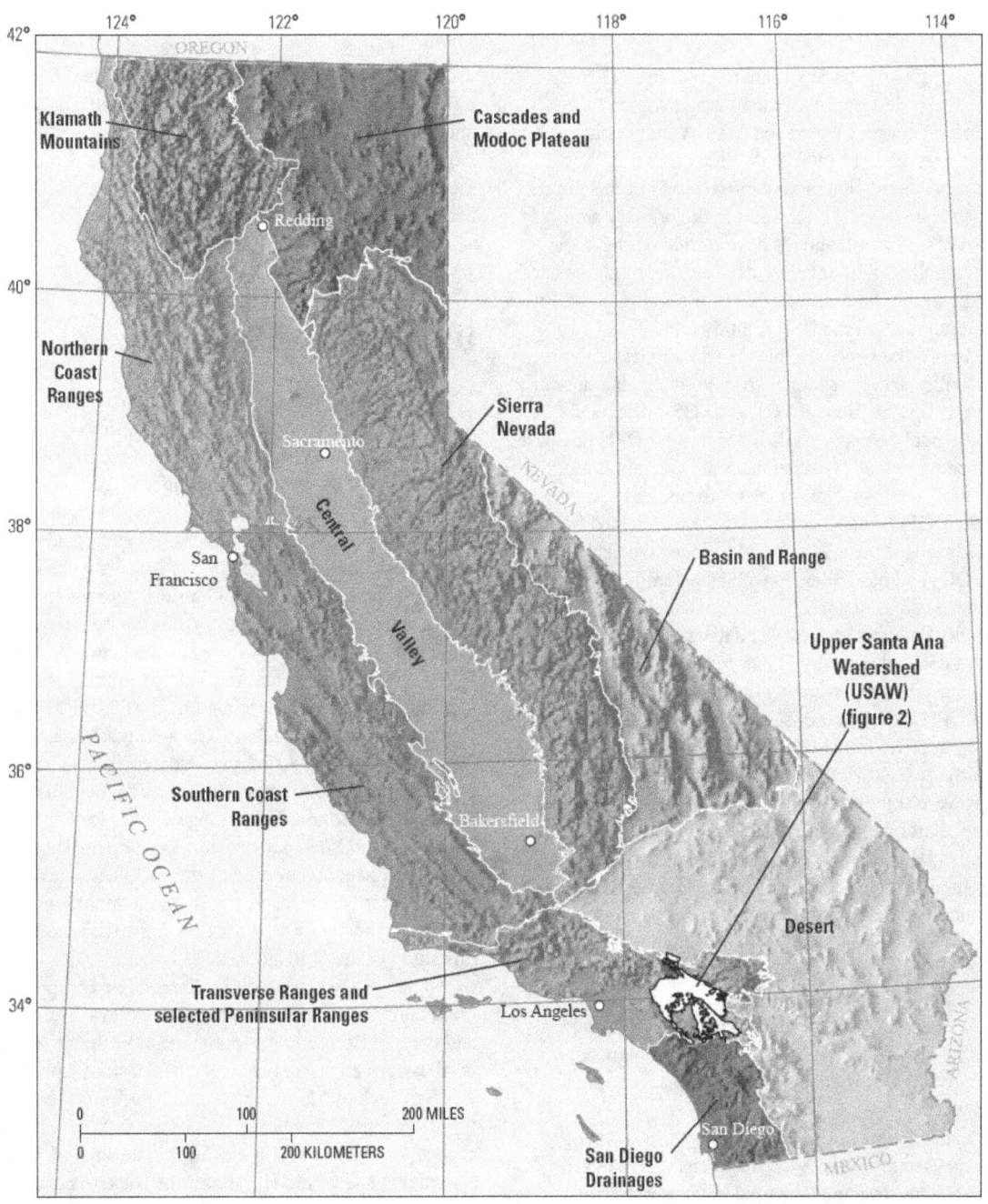

Shaded relief derived from U.S. Geological Survey
National Elevation Dataset, 2006.
Albers Equal Area Conic Projection

Provinces from Belitz and others, 2003

Figure 1.   Upper Santa Ana Watershed study unit, California GAMA Priority Basin Project, and the California hydrogeologic provinces.

Purpose and Scope

The purposes of this report are to provide a (1) *study unit description*: description of the hydrogeologic setting of the Upper Santa Ana Watershed GAMA study unit, hereinafter referred to as the USAW study unit (fig. 1), (2) *status assessment*: assessment of the status of the current (2006–2007) quality of groundwater in the primary aquifers in the USAW study unit, and (3) *compilation of ancillary datasets*: compilation of ancillary datasets that might be used to help explain the status assessment of groundwater quality in the primary aquifers in the USAW study unit.

The *status assessment* in this report includes analyses of water-quality data for samples from 90 wells (hereinafter referred to as USGS-grid wells). The USGS-grid wells mostly were public-supply wells listed in the CDPH database, but included other wells (irrigation, domestic, monitoring, and industrial) with perforation intervals similar to wells listed in the CDPH database. Samples were collected from USGS-grid wells for analysis of anthropogenic constituents such as VOCs and pesticides, as well as naturally occurring constituents, such as major ions, nutrients, and trace elements. Water-quality data from the CDPH database also were used to supplement data collected by USGS for the GAMA Program. The resulting set of water-quality data from USGS-grid and selected CDPH wells was considered to be representative of the primary aquifers in the USAW study unit.

To provide context, the water-quality data discussed in this report were compared to California and Federal regulatory and non-regulatory benchmarks for drinking water. The assessments in this report characterize the quality of untreated groundwater resources in the primary aquifers within the study unit, not the treated drinking water delivered to consumers by water purveyors. This study does not attempt to evaluate the quality of water delivered to consumers; after withdrawal from the ground, water typically is treated, disinfected, and (or) blended with other waters to maintain acceptable water quality for consumers. Regulatory benchmarks apply to drinking water that is delivered to the consumer, not to untreated groundwater.

The appendixes of this report include discussion of the methods used to attribute wells and characteristics of explanatory factors that may be used in an assessment for understanding in future reports. Potential explanatory factors examined included land use, well depth, indicators of groundwater age, pH, and oxidation-reduction conditions. In addition to the 90 grid wells sampled for the status assessment, 9 additional wells were sampled by USGS for the purpose of understanding some known or suspected water-quality issue in the study unit (hereinafter referred to as understanding wells). Attributes of all grid and understanding wells are presented in appendix A.

Water-quality data for samples collected by the USGS for the GAMA Program in the USAW study unit and details of sample collection, analysis, and quality-assurance procedures

for the USAW study unit are presented by Kent and Belitz (2009). Using the same data, this report describes methods used in designing the sampling network, identification of CDPH data for use in the *status assessment*, analysis of ancillary datasets, and estimation of aquifer-scale proportions. Aquifer-scale proportions are defined as the percentage of the area of the primary aquifer system with concentrations above or below specified thresholds relative to regulatory or aesthetic benchmarks.

Description of Study Unit

The USAW study unit covers approximately 1,000 square miles (mi^2) in Riverside and San Bernardino Counties and has a population of nearly two million people (California Department of Finance, 2000). The USAW study unit lies within the Transverse Ranges and selected Peninsular Ranges hydrogeologic province (fig. 1) and contains three groundwater basins—Upper Santa Ana Valley, San Jacinto, and Elsinore—considered high priority for assessment by Belitz and others (2003) (fig. 2). For the purpose of this study, these groundwater basins were grouped into six study areas (fig. 3). The Upper Santa Ana Valley Groundwater Basin includes nine subbasins defined by the CDWR: Bunker Hill, Cajon, Rialto-Colton, Chino, Cucamonga, Yucaipa, San Timoteo, Riverside-Arlington, and Temescal. These were combined into the following four study areas: Bunker Hill/Cajon/Rialto-Colton, Cucamonga/Chino, Riverside-Arlington/Temescal, and Yucaipa/San Timoteo. The fifth and sixth study areas were composed of the San Jacinto and Elsinore groundwater basins respectively. Mountainous areas, consisting mostly of hard rock geology, were excluded from this study of the alluvial basins.

The USAW study unit is characterized by prominent mountains that rise steeply from the valleys. The San Gabriel and the San Bernardino Mountains make up the northern and northeastern edges, respectively, of the study unit (fig. 2). The San Jacinto Mountains lie on the southeastern edge of the study unit. The tallest peaks of each of these three ranges exceed 10,000 feet (ft) in elevation. The smaller Santa Ana Mountains and Chino Hills form the western edge of the study unit, and separate it from the Coastal Santa Ana Basin. The lowest elevation in the USAW region is about 500 ft above sea level in the area around Prado Dam in Corona, and most of the study unit valley floors are less than 2,000 ft above sea level. The climate of the USAW region is Mediterranean, with hot, dry summers and cool, wet winters. Temperatures range from daytime highs of about 90 degrees Fahrenheit (°F) in summer to night-time lows of about 40°F in winter (Danskin and others, 2006). Average annual precipitation ranges from 10 to 24 inches in the valleys and from 24 to 48 inches in the mountains, where much of it comes in the form of snow (Belitz and others, 2004).

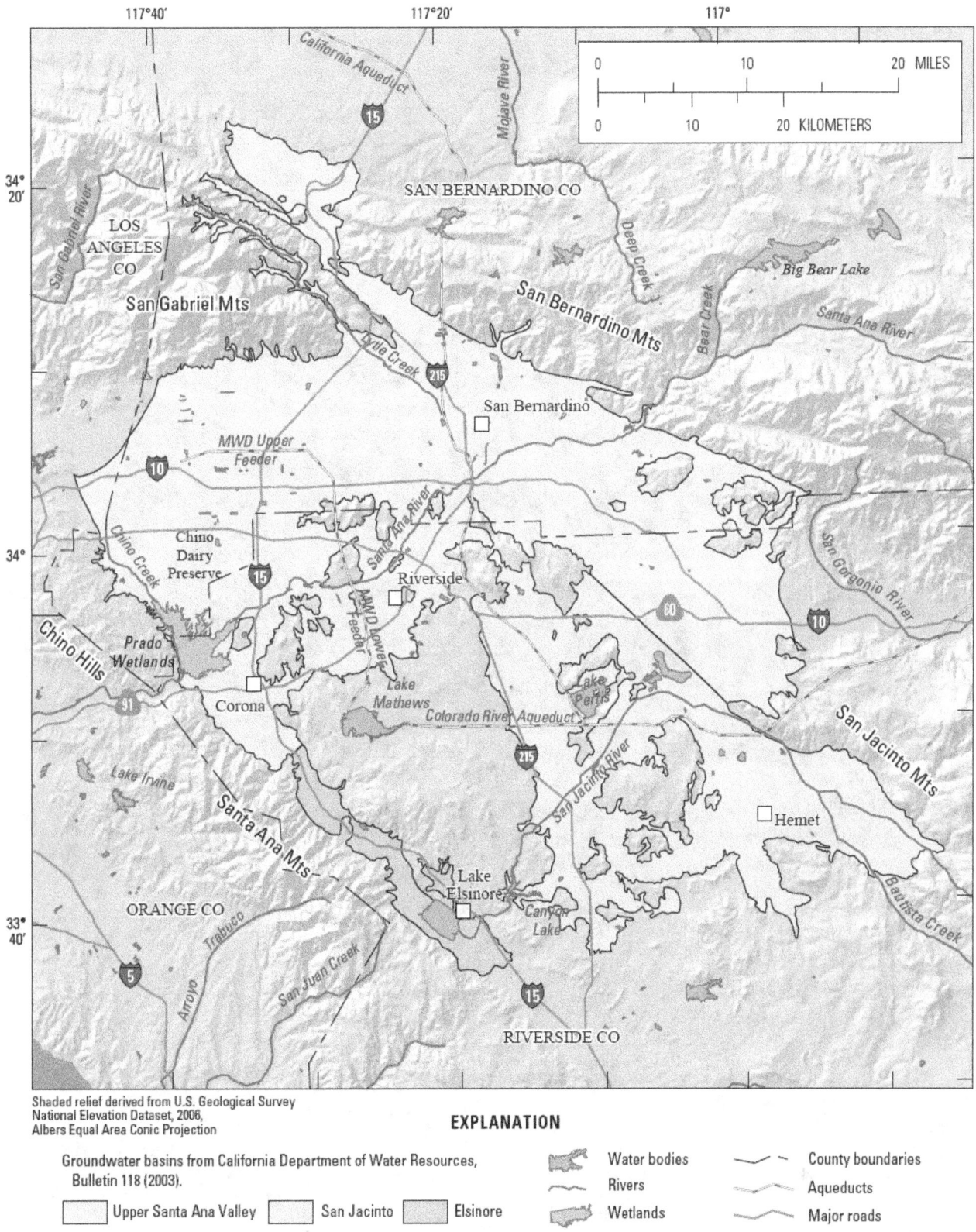

Figure 2.   Geographic and cultural features of the Upper Santa Ana Watershed study unit, California GAMA Priority Basin Project.

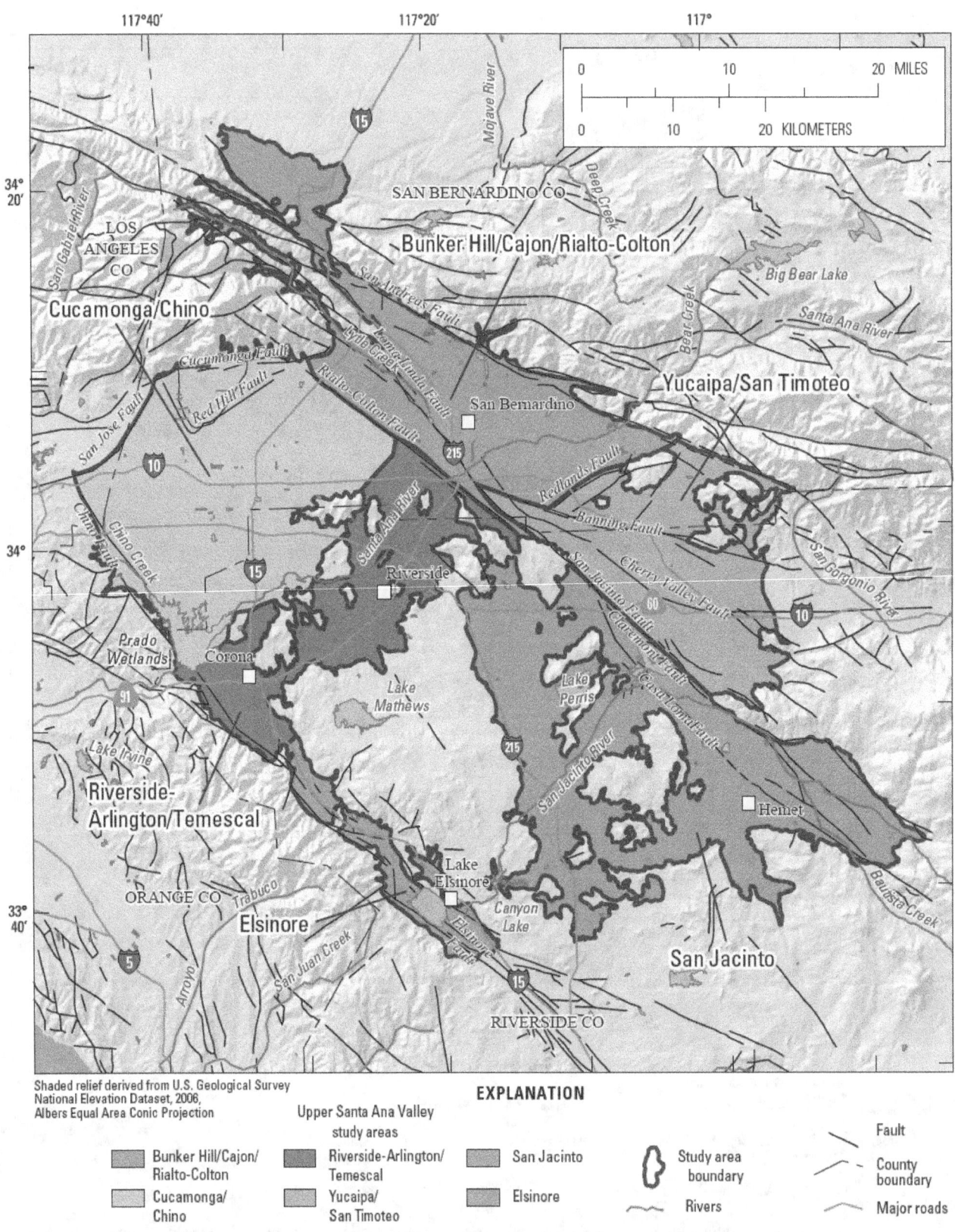

Figure 3.   Study areas of the Upper Santa Ana Watershed study unit, California GAMA Priority Basin Project.

Land Use

Land use in the study unit is 44 percent natural, 21 percent agricultural, and 35 percent urban, based on classifications from USGS National Land Cover Data (Nakagaki and others, 2007) (figs. 4, D1A). Agricultural land use in USAW is mostly in the San Jacinto study area and the Chino Dairy Preserve portion (fig. 2) of the Cucamonga-Chino study area. However, the area of the Chino Dairy Preserve is being converted rapidly to an urban area, and most of the rest of the Cucamonga-Chino study area is urban (fig. 4). The Yucaipa-San Timoteo and Elsinore study areas have predominantly natural land cover (fig. 4). Natural lands are mostly steep areas that are difficult to develop, or forests on the edges of the study unit. Most of the land use adjacent to the USAW study unit is natural and consists of steep mountains or hills generally covered by forest or chaparral.

Hydrogeologic Setting

Aquifers of the Upper Santa Ana Valley Groundwater Basin are generally unconfined and consist of alluvial deposits eroded from the surrounding mountains filling several subbasins (fig. 5) (Hamlin and others, 2005; California Department of Water Resources, 2004a,b,c,d,e,f,g; 2006a,b,c,d). The thicknesses of these alluvial deposits range from less than 200 ft to more than 1,000 ft (Dutcher and Garrett, 1963). Faults play an important role in the groundwater flow system. The San Andreas Fault, which lies along the base of the San Bernardino Mountains, and other faults, which lie along the base of the San Gabriel Mountains and Chino Hills, bound the groundwater basin on three sides (fig. 3) (Hamlin and others, 2002). Other faults, such as the Rialto-Colton Fault, divide the Upper Santa Ana Valley Groundwater Basin into its subbasins. These interior faults locally influence groundwater flow and control the location of groundwater discharge (Woolfenden and Kadhim, 1997; Izbicki and others, 1998; Hamlin and others, 2002). Groundwater flow in the San Bernardino area of the Upper Santa Ana Valley, known as the Bunker Hill groundwater subbasin, is characterized by flow paths that originate along the mountain front and converge to a focused discharge area in San Bernardino near the convergence of the Santa Ana River and the San Jacinto Fault (Wildermuth Environmental, Inc., 2000; Dawson and others, 2003).

Groundwater flow in aquifers of the Elsinore Basin is also affected by several faults cutting alluvial and lacustrine sediments (fig. 5) (California Department of Water Resources, 2006c). Flood-plain deposits in the interior of the valley typically reach a thickness of about 200 ft, while the principal water-bearing unit of the basin beneath Lake Elsinore reaches a maximum thickness of 2,200 ft (California Department of Water Resources, Southern District, 1981; California Department of Water Resources, 2006c).

Aquifers of the San Jacinto Groundwater Basin are generally unconfined and consist of a series of interconnected alluvium-filled valleys bounded by steep-sided bedrock mountains and hills (fig. 5) (Hamlin and others, 2005). However, some confined conditions occur in the eastern part of the basin (California Department of Water Resources, 2006d). The deposits in the San Jacinto Basin valleys typically are 200 to 1,000 ft thick (Eastern Municipal Water District, 2002), but may exceed 5,000 ft in the eastern part of the basin between the Casa Loma and Claremont Faults (fig. 3) (California Department of Water Resources, 2006d).

Water Management

Two stipulated judgments broadly adjudicated water rights in the Santa Ana Basin in 1969 (Orange County Water District v. City of Chino, Superior Court No. 117628; and Western Municipal Water District of Riverside County v. East San Bernardino County Water District, Superior Court No. 78426). Currently, the Santa Ana Watermaster compiles hydrologic and water-quality data in annual reports, and the Santa Ana Watershed Project Authority, a Joint Powers Authority, classified as a Special District (government agency), plans and builds facilities to protect the water quality of the Santa Ana River Watershed. In addition, the Chino Basin was separately adjudicated in 1978; the Chino Basin Watermaster was directed to establish a comprehensive basin management program there (Miller and others, 2007). Partly as a result of these cases, water managers in the USAW study unit have been proactive in their response to declines in the quantity and quality of groundwater.

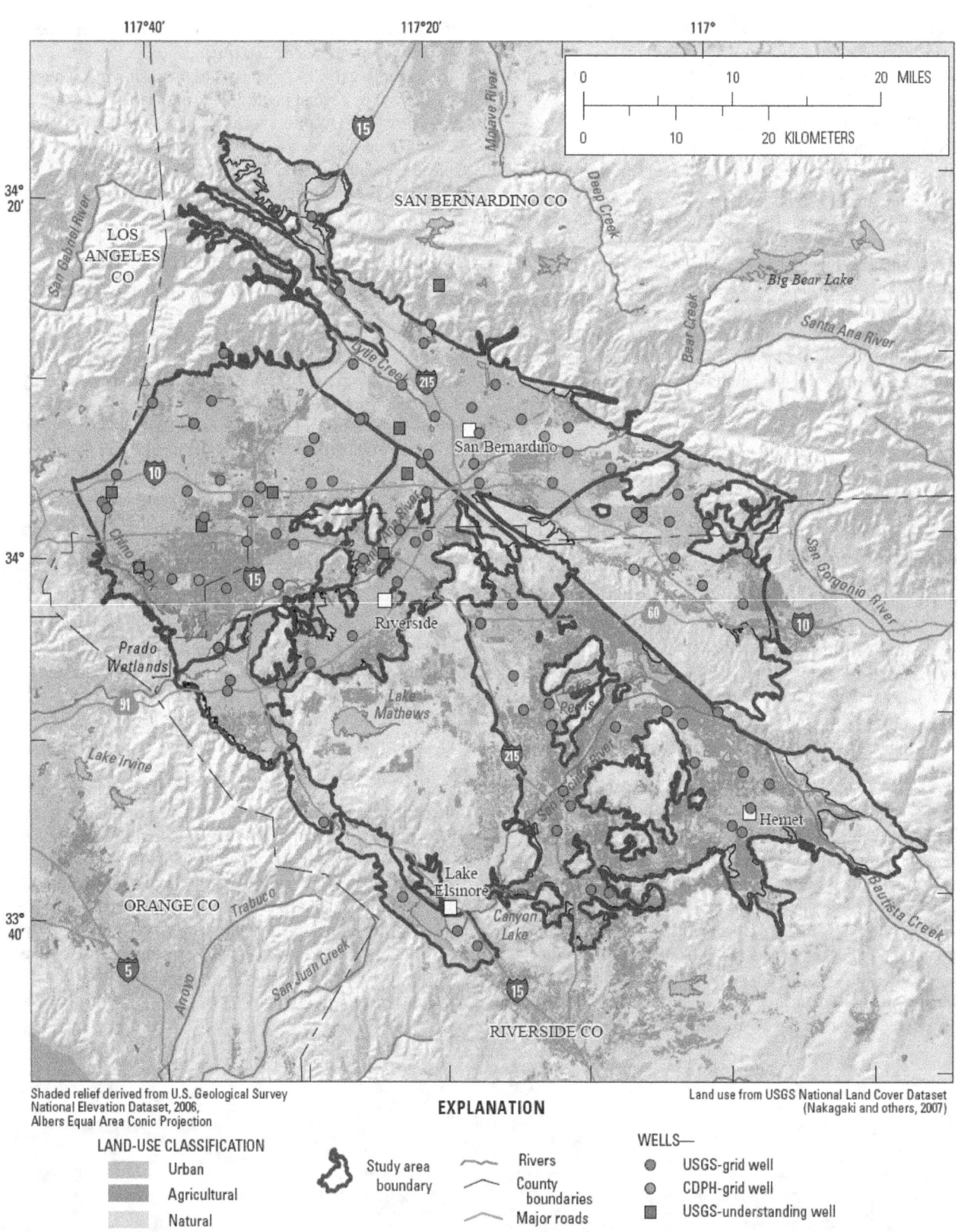

EXPLANATION

Land use from USGS National Land Cover Dataset
(Nakagaki and others, 2007)

LAND-USE CLASSIFICATION

Urban

Agricultural

Natural

Study area
boundary

Rivers

County
boundaries

Major roads

WELLS—

USGS-grid well

CDPH-grid well

USGS-understanding well

Figure 4.   Land use, locations of grid and understanding wells sampled, and locations of California Department of Public Health wells used for supplemental data on inorganic constituents, Upper Santa Ana Watershed study unit, California GAMA Priority Basin Project.

Shaded relief derived from U.S. Geological Survey
National Elevation Dataset, 2006,
Albers Equal Area Conic Projection

EXPLANATION

Geologic unit

- Quaternary alluvium
- Quaternary other sediment
- Plio-Pleistocene sediment
- Quarternary, mafic volcanics
- Tertiary nonmarine sediment

- Tertiary marine sediment
- Tertiary, mafic volcanics
- Franciscan complex
- Pyroclastic volcanics

- Granitic rocks
- Metamorphic other
- Metasediment
- Ultramafic/mafic

- Study unit and area boundary
- Fault—Dashed where approximately located, dotted where concealed, queried where uncertain
- Water bodies
- Rivers
- Wetlands

Figure 5. Geologic formations of the Upper Santa Ana Watershed study unit, California GAMA Priority Basin Project.

Methods

Methods used for the GAMA Priority Basin Project were selected to achieve the following objectives: (1) design a sampling plan suitable for statistical analysis, (2) combine CDPH data with data collected in 2006–07 by the USGS for assessing water quality, (3) evaluate proportions of the primary aquifers having high, moderate, and low concentrations for constituent classes and individual constituents for additional evaluation, (4) select constituents for additional evaluation, and (5) compile and classify relevant ancillary data, so that relations of potential explanatory factors to water quality might be identified and discussed in future reports.

This study was designed to provide a spatially unbiased assessment of untreated groundwater quality within the primary aquifer systems. The primary metric for defining groundwater quality in this study was *relative-concentration*, which compares groundwater chemistry to regulatory and non-regulatory benchmarks used to evaluate drinking-water quality. All constituents with benchmarks were included in the status assessment. Constituents were selected for additional evaluation in the assessment on the basis of objective criteria by using their measured relative-concentrations. Groundwater-quality data collected by the USGS for the GAMA Program and data compiled in the CDPH database were used in the *status assessment*. Two statistical approaches based on spatially unbiased grids with equal-area cells within each study area were used to calculate aquifer-scale proportions of the three relative-concentration categories.

The *status assessment* included two primary steps. First, water-quality data were normalized to their respective water-quality benchmarks by calculating the relative-concentrations of constituents (Toccalino and others, 2004; Toccalino and Norman, 2006). Second, aquifer-scale proportions were determined for categories of "high," "moderate," and "low" based on the spatial aggregation of the relative-concentrations using two approaches: (1) grid based, and (2) spatially weighted. The grid-based approach uses one well per cell to represent groundwater quality, and water-quality data are from wells sampled by the USGS, augmented with data from selected wells in the CDPH database. The spatially weighted approach uses data for wells sampled by the USGS and all wells in the CDPH database, and weights the relative-concentration category (high, moderate, low) of each well such that each grid cell contributes equally to represent groundwater quality. The influence (weight) of each well's relative-concentration category is reduced in proportion to the number of wells in its cell. In turn, the influence of each cell to determine aquifer proportion in the study unit is reduced in proportion to the number of cells in the study unit. Results for the two approaches were compared, and results from the preferred approach were used to select constituents for additional evaluation.

Relative-Concentrations and Water-Quality Benchmarks

Concentrations of water-quality constituents are presented as relative-concentrations in the *status assessment*, where

$$\text{Relative-concentration} = \frac{\text{Sample concentration}}{\text{Water-quality benchmark concentration}}.$$

Relative-concentrations provide a means to relate concentrations of constituents in groundwater samples to water-quality benchmarks. Relative-concentrations less than 1.0 indicate sample concentrations less than the benchmark; relative-concentrations greater than 1.0 indicate sample concentrations greater than the benchmark. The use of relative-concentrations also normalizes a wide range of concentrations for different constituents to a common scale relative to benchmarks. Toccalino and others (2004), Toccalino and Norman (2006), and Rowe and others (2007) previously used the ratio of measured concentration to a benchmark [either maximum contaminant levels (MCLs) or Health-Based Screening Levels (HBSLs)] and defined this ratio as the *benchmark quotient*. Relative-concentrations used in this report are equivalent to the benchmark quotient reported by Toccalino and others (2004) for constituents that have water-quality benchmarks. HBSLs were not used in this report because they are not currently used as benchmarks by California drinking-water regulatory agencies. Relative-concentrations were only computed for compounds with water-quality benchmarks. About half of the constituents analyzed for in the USAW study unit have benchmarks.

Regulatory and non-regulatory water-quality benchmarks apply to water that is served to the consumer, not to untreated groundwater. However, to provide context for the water-quality results, concentrations of constituents measured in the untreated groundwater were compared with regulatory and non-regulatory human-health-based water-quality benchmarks established by the U.S. Environmental Protection Agency (USEPA) and CDPH (U.S. Environmental Protection Agency, 2006; California Department of Public Health, 2008a). The human-health benchmarks used include MCLs, notification levels (NLs), health advisory levels (HALs), action levels (ALs), and risk-specific dose (1 in 100,000 lifetime risk of cancer; RSD5-US). Non-regulatory benchmarks set for aesthetic concerns, secondary maximum contaminant levels defined by CDPH and USEPA (SMCL-CA and SMCL-US, respectively), also were used. If a constituent had multiple types of benchmarks, the benchmark used for calculation of relative-concentration was selected according to the following order of priority: regulatory human-health (MCL and AL), non-regulatory aesthetic (SMCL), and non-regulatory human-health (NL-CA, HAL-US, and RSD5-US, in that order). For the regulatory human-health benchmarks, Federal benchmark levels were used, unless the California levels were lower. California public health goals were not used in this report. Additional information on the types of benchmarks and the benchmarks for all constituents analyzed are provided by Kent and Belitz (2009).

Relative-concentrations were classified into high, moderate, and low categories.

Category	Relative-concentrations for organic and special-interest constituents	Relative-concentrations for inorganic constituents
High	> 1	> 1
Moderate	> 0.1 and $\leq$ 1	> 0.5 and $\leq$ 1
Low	$\leq$ 0.1	$\leq$ 0.5

A relative-concentration of 0.1 was used as a threshold between low and moderate values of organic and special-interest constituents compared with a relative-concentration of 0.5 for inorganic constituents. A larger threshold value was used for inorganic constituents because naturally occurring inorganic constituents tend to be more prevalent than organic constituents in California groundwater (Landon and others, 2010). Also, the USEPA has used a relative-concentration of 0.1 of the regulatory benchmark as a threshold concentration at or above which the USEPA wants to be informed of a pesticide's presence in surface water or groundwater (U.S. Environmental Protection Agency, 1998). In contrast, inorganic constituents typically occur naturally at concentrations that could be greater than 0.1 of regulatory benchmarks; consequently, it would be difficult to identify the highest-priority inorganic constituents that may have elevated concentrations above background levels if a relative-concentration of 0.1 were used as the threshold between moderate and low relative-concentrations. Therefore, a relative-concentration of 0.5 was used as a threshold between low and moderate values of inorganic constituents, rather than 0.1 as was used for the organic and special-interest constituents.

Design of Sampling Network for Status Assessment

The wells selected for sampling by the USGS in this study were selected to provide a statistically unbiased, spatially distributed set of wells for the assessment of the quality of groundwater in the primary aquifers. Water-quality data from the USGS-grid wells were augmented with data from selected wells from the CDPH database (CDPH-grid wells, discussed further in next section) to obtain more complete grid coverage, including constituents that were not analyzed for at every USGS-grid well. These data were used to assess proportions of the primary aquifer system having high, moderate, and low relative-concentrations.

The primary data used for the grid-based calculations of aquifer-scale proportions were data from wells sampled by the GAMA Priority Basin Project. Detailed descriptions of the methods used to identify wells for sampling are given in Kent and Belitz (2009). Briefly, each study area was divided into equal-area grid cells, and in each cell, one well was randomly selected to represent the cell (figs. 6, A1) (Scott, 1990). Wells were selected from the population of wells in statewide databases maintained by the CDPH and the USGS. Water-quality data from the USGS-grid wells were combined with data from selected wells from the CDPH database (CDPH-grid wells) to provide better spatial grid coverage (fig. A2), including data for constituents not analyzed for at every USGS-grid well, to assess proportions of the primary aquifers having high, moderate, and low relative-concentrations. In addition, nine understanding wells were selected for sampling by the USGS to increase the density in several areas to address specific groundwater-quality issues in the study unit.

The USGS-grid wells were selected by using a randomized grid-based method (Scott, 1990). The network of grid wells was selected by first plotting the location of wells listed in the statewide databases maintained by the CDPH and USGS on a regional map of the six study areas. Five of the six study areas were divided into grids of equal-area cells (10 mi^2; ~26 km^2)—the Bunker Hill/Cajon/Rialto-Colton (USAWB) study area (20 grid cells), the Yucaipa/San Timoteo (USAWY) study area (12 grid cells), the Riverside-Arlington/Temescal (USAWR) study area (13 grid cells), the Cucamonga/Chino (USAWC) study area (27 grid cells), and the San Jacinto (USAWS) study area (31 grid cells) (fig. 6). One public-supply well per 10-mi^2 grid cell was randomly chosen to be sampled (Kent and Belitz, 2009). The relatively small (approximately 40 mi^2) Elsinore groundwater basin (USAWE) has an uneven distribution of available wells to sample and was not divided into cells; instead, four wells (approximately one well per 10 mi^2) that were spread throughout the basin were chosen for sampling to represent four "equivalent cells." The varied shapes of the equal-area grid cells were drawn by objectively using the method reported by Scott (1990) and were influenced by the irregular shapes of the boundaries of the study areas (figs. 6, A1, A2). Geographic features near the edges of some study areas caused some grid cells to be divided into more than one section to achieve a 10-mi^2 area for these cells. If a grid cell contained more than one public-supply well, each well was randomly assigned a rank. The highest ranking well that met basic sampling criteria (for example, sampling point located upstream of treatment, capability to pump for several hours, and available well construction information), and for

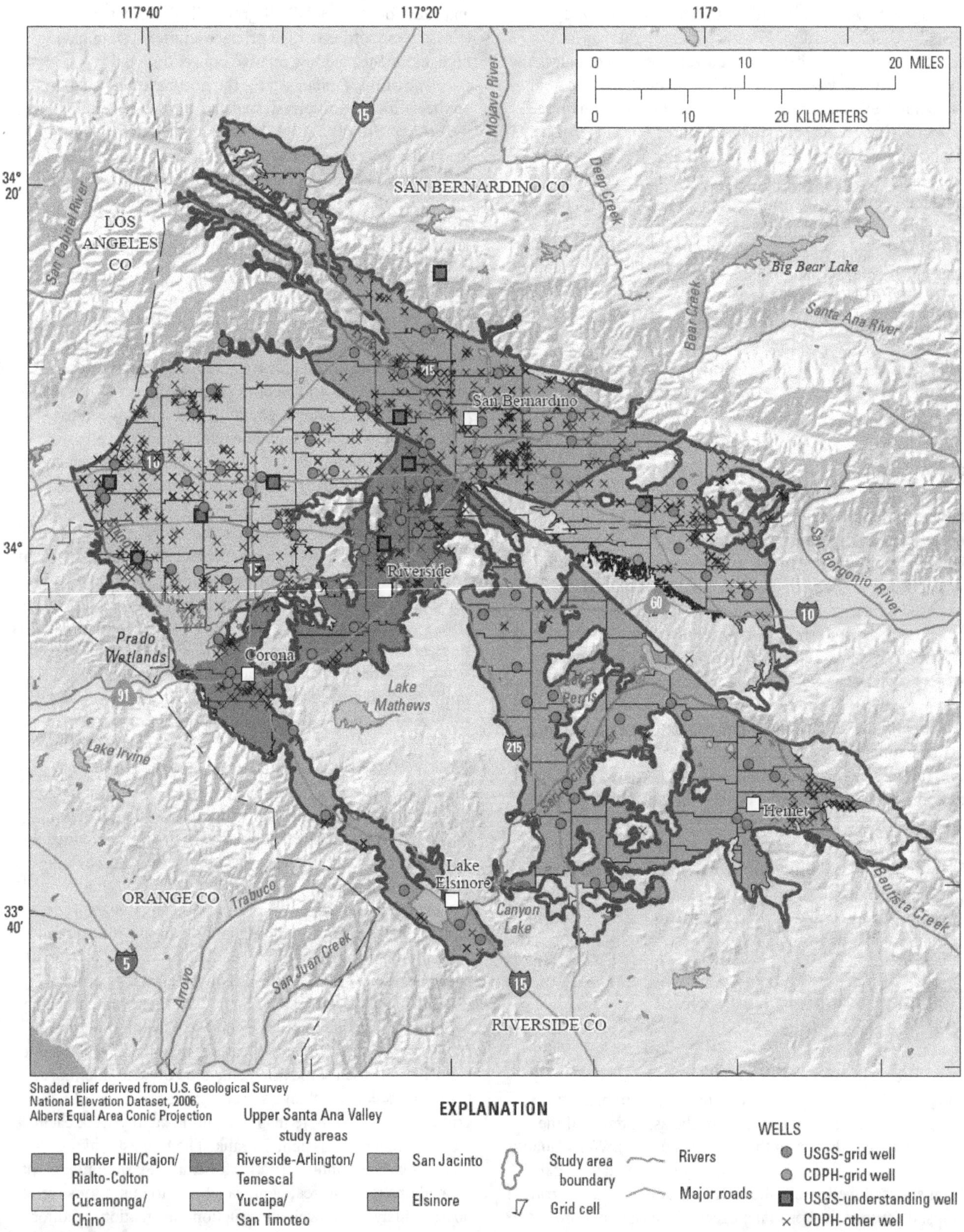

Figure 6.   Grid cells for each study area, grid and understanding wells sampled during November 2006–March 2007, grid wells from which data for inorganic constituents from the California Department of Public Health (CDPH) database were used, and all CDPH wells in the Upper Santa Ana Watershed study unit, California GAMA Priority Basin Project.

which permission to sample could be obtained, was selected. If a grid cell did not contain accessible public-supply wells, then irrigation, monitoring, domestic, or other types of wells were considered. The USGS-grid wells were sampled by the USGS for the Priority Basins Project, but are owned by other organizations or individuals. Grid wells in the USAW study unit were labeled with prefixes that vary by study area (USAWB-, USAWC-, USAWE-, USAWR-, USAWS-, or USAWY-) (table A1).

USGS-grid wells were sampled in 19 of the 20 grid cells in the Bunker Hill/Cajon/Rialto-Colton study area, 9 of the 12 grid cells in the Yucaipa/San Timoteo study area, 12 of the 13 grid cells in the Riverside-Arlington/Temescal study area, 25 of the 27 cells in the Cucamonga/Chino study area, and 21 of the 31 grid cells in the San Jacinto study area (fig. 6). Seventeen grid cells were not sampled because they either had no wells, or permission to sample was not granted. As previously stated, four wells were sampled in the Elsinore basin study area, and these are considered grid wells representing the "equivalent cells" for the purposes of statistical analyses. The 90 USGS-grid wells sampled included 73 public supply, 8 irrigation, 5 desalter (wells that extract groundwater with high salinity for treatment), 1 monitoring, 1 domestic, 1 industrial, and 1 recreation well (used exclusively to maintain water hazards at a golf course). The seventeen grid wells that were not public supply wells had screened intervals at depths similar to those of the public supply grid wells.

The grid wells in USAW were sampled by using a tiered analytical approach (Kent and Belitz, 2009). All wells were sampled for a standard set of constituents, including water-quality indicators (field parameters), VOCs, pesticides, perchlorate, pharmaceutical compounds, noble gases, and the stable isotopes of hydrogen and oxygen in water (table 1). The standard set of constituents was termed the "fast" schedule. Forty-one grid wells and 1 understanding well were sampled on the fast schedule. Wells on the "intermediate" schedule were sampled for all of the constituents on the fast schedule, plus inorganic constituents and selected hydrologic tracers. Sixteen grid wells and 2 understanding wells were sampled for the intermediate schedule. The "topical" schedule was the same as the intermediate schedule, except that it did not include pesticide or pharmaceutical compounds. One grid well and six understanding wells were sampled on the topical schedule (the grid well was designated as an understanding well at the time that it was sampled). Wells on the "slow" schedule were sampled for all of the constituents

on the intermediate schedule, plus dissolved organic carbon, 1,4-dioxane, N-nitrosodimethylamine (NDMA), low-level 1,2,3-trichloropropane (1,2,3-TCP), the species of arsenic, chromium, and iron, an additional analysis for tritium, and radioactive and microbial constituents (table 1). Thirty-two grid wells were sampled on the slow schedule. The collection, analysis, and quality-assurance of the samples are described by Kent and Belitz (2009).

CDPH-Grid Well Selection

Samples to be analyzed for inorganic constituents were collected by the USGS from 49 of the 107 grid cells. The CDPH database was used to provide data for inorganic constituents for the cells that lacked these data (table 2). In this way, at least some inorganic data were obtained for an additional 34 grid cells. For 29 of these 34 cells, the inorganic data were obtained for the same grid well sampled by the USGS. In these cases, the grid well identifier contains "DG" (fig. A2; table A1). CDPH inorganic data for three of the additional grid cells were obtained for cells sampled by the USGS, but at wells other than the ones sampled by the USGS. In these cases, the grid well identifier contains "DPH" (fig. A2; table A1). In addition, CDPH inorganic data were obtained for two grid cells that were not sampled by the USGS. In these two cases, the grid well identifier also contains "DPH" (USAWC-DPH-1 and USAWY-DPH-1; fig. A2; table A1). No CDPH inorganic data were available for nine grid cells sampled by the USGS for organic constituents. Finally, 15 of the 107 grid cells had no wells available for sampling and no wells with data in the CDPH database (table A1).

The CDPH database generally did not contain data for all missing inorganic constituents at every CDPH grid well; therefore, the number of wells used for the grid-based assessment differed for various inorganic constituents (table 2). Although other organizations also collect water-quality data, the CDPH data is the only statewide database of groundwater-chemistry data available for comprehensive analysis. All other CDPH wells with data from the current period (November 30, 2003, through December 1, 2006) not selected to be CDPH-grid wells are referred to as "CDPH-other" wells. Data from these "CDPH-other" wells were used to calculate raw detection frequencies of water-quality constituents.

Table 1.   Constituent class and the number of constituents and wells sampled for each analytical schedule in the Upper Santa Ana Watershed study unit, California GAMA Priority Basin Project, November 2006–March 2007.

[1,2,3-TCP, 1,2,3-trichloropropane; NDMA, *N*-nitrosodimethylamine; TDS, total dissolved solids]

	Analytical schedule			
	Fast	**Intermediate**	**Topical**	**Slow**
Well summary	**Number of wells**			
Total number of wells	42	18	7	32
Number of grid wells sampled	41	16	1	32
Number of understanding wells sampled	1	2	6	0
Analyte Groups	**Number of constituents**			
Field measurements: specific conductance, pH, dissolved oxygen, and temperature	4	4	4	4
Volatile organic compounds (VOCs) [1]	84	84	84	84
Pesticides and pesticide degradates	83	83		83
Polar pesticides and degradates [2]	52	52		52
Perchlorate	1	1	1	1
Noble gases [3]	6	6	6	6
Stable isotopes of hydrogen and oxygen of water	2	2	2	2
Pharmaceutical compounds	14	14		14
Alkalinity [4]		1	1	1
Major and minor ions, silica, TDS, and trace elements		37	37	37
Dissolved gases (carbon dioxide, argon, methane, nitrogen, oxygen) [5]		3	3	3
Carbon isotopes		2	2	2
Isotopes of nitrogen and oxygen in nitrate		2	2	2
Isotopes of nitrogen gas		1	1	1
Nutrients		5	5	5
Dissolved organic carbon				1
1,4-Dioxane, NDMA, and low-level 1,2,3-TCP [6]				3
Arsenic, chromium, and iron species [7]				6
Tritium [8]				1
Radon, uranium, and radium isotopes				4
Gross alpha and beta radioactivity				2
Microbial constituents				4
Total:	246	298	149	318

[1] Includes nine constituents classified as fumigants.

[2] Excludes six constituents in common with first group of "pesticides and pesticide degradates," as well as caffeine, counted with pharmaceutical compounds below.

[3] Analyzed at Lawrence Livermore National Laboratory (LLNL). Samples were also submitted to LLNL for tritium analysis, but LLNL tritium results are not available.

[4] Lab alkalinities only for intermediate and topical samples. Both field and lab alkalinities for slow samples.

[5] Argon analyzed in all samples as a noble gas. Dissolved oxygen measured in all samples as a field parameter.

[6] 1,2,3-TCP analyzed with a method reporting level of 0.005 microgram per liter (µg/L) here was also on the USGS VOC analytical schedule with a laboratory reporting level of 0.12 µg/L.

[7] Total dissolved arsenic, chromium, and iron results are counted with trace elements above.

[8] Analyzed at U.S. Geological Survey Tritium Laboratory, Menlo Park, California.

Table 2. Inorganic constituents and associated benchmark information, and the number of grid wells with U.S. Geological Survey-GAMA data and CDPH data, for each constituent, Upper Santa Ana Watershed study unit, California GAMA Priority Basin Project.

[CDPH, California Department of Public Health; MCL-CA, CDPH maximum contaminant level; MCL-US, U.S. Environmental Protection Agency maximum contaminant level; SMCL, secondary maximum contaminant level; NL, notification level; AL, action level; HAL, lifetime health advisory level; mg/L, milligrams per liter; µg/L, micrograms per liter; pCi/L, picocuries per liter; USGS, U.S. Geological Survey]

Constituent	Benchmark type	Benchmark value	Units	Number of grid wells with USGS-GAMA data	Number of grid wells with CDPH data
Nutrient					
Ammonia, as nitrogen	HAL-US	30	mg/L	49	0
Nitrate plus nitrite, as nitrogen	MCL-US	10	mg/L	49	34
Nitrite, as nitrogen	MCL-US	1	mg/L	49	30
Trace element					
Aluminum	MCL-CA	1,000	µg/L	49	30
Antimony	MCL-US	6	µg/L	49	30
Arsenic	MCL-US	10	µg/L	49	30
Barium	MCL-CA	1,000	µg/L	49	30
Beryllium	MCL-US	4	µg/L	49	30
Boron	NL-CA	1,000	µg/L	49	18
Cadmium	MCL-US	5	µg/L	49	30
Chromium	MCL-CA	50	µg/L	49	30
Copper	AL-US	1,300	µg/L	49	32
Iron	SMCL-CA	300	µg/L	49	32
Lead	AL-US	15	µg/L	49	29
Manganese	SMCL-CA	50	µg/L	49	32
Mercury	MCL-US	2	µg/L	32	31
Molybdenum	HAL-US	40	µg/L	49	0
Nickel	MCL-CA	100	µg/L	49	30
Selenium	MCL-US	50	µg/L	49	30
Silver	SMCL-CA	100	µg/L	49	30
Strontium	HAL-US	4,000	µg/L	49	0
Thallium	MCL-US	2	µg/L	49	30
Vanadium	NL-CA	50	µg/L	49	13
Zinc	SMCL-US	5,000	µg/L	49	32
Minor ion					
Fluoride	MCL-CA	2	mg/L	49	30
Major ion					
Chloride	SMCL-CA	500	mg/L	49	32
Sulfate	SMCL-CA	500	mg/L	49	32
Total dissolved solids (TDS)	SMCL-CA	1,000	mg/L	49	32
Radioactive					
Gross alpha radioactivity, 72-hour count	MCL-US	15	pCi/L	32	22
Gross beta radioactivity, 72-hour count	MCL-CA	50	pCi/L	32	0
Radium-226 + Radium-228	MCL-US	5	pCi/L	32	11
Radon-222	MCL-US [1]	4,000	pCi/L	32	4
Uranium	MCL-US	20	pCi/L	49	12

[1] Proposed.

Samples from all 90 USGS-grid wells were analyzed for VOCs and perchlorate. Samples from 89 of these USGS-grid wells were analyzed for pesticide compounds (only USAWC-24 lacks USGS pesticide data). VOC and perchlorate data are lacking for 27 out of 107 total grid cells; pesticide data are lacking for 28 grid cells. The USGS-grid-well data for VOCs, perchlorate, and pesticides have lower reporting levels than are available from the CDPH database (table 3); therefore, CDPH data for these constituents were not used to supplement USGS-grid-well data for the status assessment.

The CDPH database contains more than 1.5 million historical water-quality results from more than 1,000 wells in the USAW study unit, necessitating targeted retrievals to effectively access water-quality data. CDPH data were used with USGS-grid data to identify constituents in the USAW study unit with concentrations greater than water-quality benchmarks at any time during the period of record. Data were retrieved from the CDPH database for samples from all wells located within the USAW study unit for the full period of record (July 13, 1956, to December 1, 2006). Concentrations of constituents were identified as "historically high" (table 4) if they had high relative-concentrations at any time before November 30, 2003, and during the period of record, but did not have high relative-concentrations in the most recent 3-year period of CDPH data (November 30, 2003, through December 31, 2006, hereinafter referred to as current period) or in the USGS-grid data. These "historically high" constituents do not reflect current conditions on which the status assessment is based.

Table 3.   Comparison of the number of compounds and median laboratory reporting levels or method detection levels by type of constituent for data stored in the California Department of Public Health database and data collected for the Upper Santa Ana Watershed study unit, California GAMA Priority Basin Project, November 2006–March 2007.

[CDPH, California Department of Public Health; MDL, method detection limit; LRL, laboratory reporting level; mg/L, milligrams per liter; µg/L, micrograms per liter; pCi/L, picocuries per liter; SSMDC, sample-specific minimum detectable concentration; nc, not collected]

Constituent type	CDPH		GAMA		
	Number of compounds	Median MDL	Number of compounds	Median LRL	Median units
Volatile organic compounds	78	0.5	84	0.07	µg/L
Pesticides plus degradates	154	1	135	0.03	µg/L
Nutrients, major and minor ions	16	0.4	17	0.03	mg/L
Trace elements	20	10	24	0.11	µg/L
Radioactive constituents (SSMDC)[1]	8	1	9	1	pCi/L
Perchlorate	1	4	1	0.5	µg/L
N-Nitrosodimethylamine (NDMA)	1	[2]1.9	1	0.002	µg/L
Pharmaceutical compounds (MDL)	nc	nc	14	0.04	µg/L

[1] Value reported in GAMA column of median LRL is a median SSMDC for eight radioactive constituents collected and analyzed by GAMA, and excludes uranium, which had an LRL expressed as 0.04 µg/L with no SSMDC.

[2] Two detects reported by CDPH were at levels of 0.018 µg/L and 0.006 µg/L, implying a lower detection capability than given here.

Table 4. Constituents reported at concentrations greater than benchmarks in the California Department of Public Health (CDPH) database between July 13, 1956, and November 30, 2003, Upper Santa Ana Watershed study unit, California GAMA Priority Basin Project.

[high, concentration above human-heath benchmark; MCL-US, U.S. Environmental Protection Agency maximum contaminant level; MCL-CA, CDPH maximum contaminant level; NL-CA, CDPH notification level; HAL-US, U.S. Environmental Protection Agency Lifetime Health Advisory; AL-US, U.S. Environmental Protection Agency action level; mg/L, milligrams per liter; μg/L, micrograms per liter; pCi/L, picocuries per liter]

Constituent	Benchmark type	Benchmark value	Units	Date of most recent high value	Number of historically high wells	Number of wells with analysis
Trace elements						
Antimony	MCL-US	6	μg/L	10-03-02	3	647
Cadmium	MCL-US	5	μg/L	11-02-95	8	726
Copper	AL-US	1,300	μg/L	07-17-97	1	732
Mercury	MCL-US	2	μg/L	06-19-98	6	725
Thallium	MCL-US	2	μg/L	04-08-03	1	647
Nutrient						
Nitrite (as nitrogen)	MCL-US	1	mg/L	06-12-01	1	650
Radioactive constituent						
Radium-226	MCL-US	5	pCi/L	01-26-00	3	170
Trihalomethanes						
Total trihalomethanes	MCL-US[1]	80	μg/L	04-14-98	1	743
Solvents						
1,1-Dichloroethane	MCL-CA	5	μg/L	07-02-96	2	769
cis-1,2-Dichloroethene	MCL-CA	6	μg/L	05-17-91	1	733
Dichloromethane (methylene chloride)	MCL-US	5	μg/L	12-08-89	5	769
1,1,2-Trichloroethane	MCL-US	5	μg/L	06-13-86	1	769
Organic synthesis						
Chloromethane	HAL-US	30	μg/L	12-03-99	1	769
Vinyl chloride	MCL-CA	0.5	μg/L	06-11-86	2	769
Fumigants						
1,2-Dichloropropane	MCL-US	5	μg/L	09-28-98	4	763
1,2-Dibromoethane (EDB)	MCL-US	0.05	μg/L	12-19-96	3	712
Gasoline oxygenate degradate						
tert-Butyl alcohol	NL-CA	12	μg/L	04-27-01	1	553
Pesticides						
Heptachlor	MCL-CA	0.01	μg/L	11-20-89	3	627
Potential wastewater indicator						
Bis(2-ethylhexyl)phthalate	MCL-CA	4	μg/L	03-31-92	7	571
Constituent of special interest						
N-Nitrosodimethylamine (NDMA)	NL-CA	0.01	μg/L	08-01-01	1	86

[1] The MCL-US benchmark for trihalomethanes is the sum of chloroform, bromoform, bromodichloromethane, and dibromochloromethane.

Selection of Constituents for Additional Evaluation

More than 300 constituents were analyzed in samples from the USAW study unit wells; however, only a subset of these constituents is discussed in this report. Three criteria were used to select constituents for additional evaluation:

1. Constituents that were present at high or moderate relative-concentrations in the CDPH database within the 3-year interval (November 30, 2003–December 1, 2006);

2. Constituents present at high or moderate relative-concentrations in the USGS-grid wells or understanding wells; or

3. Organic constituents with detection frequencies of greater than 10 percent in the USGS-grid-well dataset for the study unit.

Constituents that were selected for additional evaluation and that were present at high relative-concentrations in greater than 2 percent of the primary aquifers are discussed in sections of this report under individual headings named for these constituents. Constituents that were selected for additional evaluation, but that were present at high relative-concentrations in less than 2 percent of the primary aquifers, including organic constituents detected at any concentration in more than 10 percent of the primary aquifers, are not given individual headings. These are discussed in sections identified by constituent class.

Calculation of Aquifer-Scale Proportions

Aquifer-scale proportions are defined as the percentage of the area (rather than the volume) of the primary aquifer system with concentrations greater or less than specified thresholds relative to regulatory or aesthetic water-quality benchmarks. Two statistical approaches were selected to evaluate the proportions of the primary aquifers (Belitz and others, 2010) in the USAW study unit with high, moderate, or low relative-concentrations of constituents relative to benchmarks:

- Grid-based: One value per grid cell, from either a USGS-grid or CDPH-grid well, was used to represent the primary aquifer system. The proportion of the primary aquifer system with high relative-concentrations was calculated by dividing the number of grid cells represented by a high relative-concentration for a particular constituent by the total number of grid cells with data for that constituent (appendix C). Proportions of moderate and low relative-concentrations were calculated similarly. Confidence intervals for grid-based detection frequencies of high relative-concentrations were computed by using the Jeffreys interval for the

binomial distribution (Brown and others, 2001). The grid-based estimate is spatially unbiased. However, the grid-based approach may not identify constituents that are present at high relative-concentrations in small areas of the primary aquifers.

- Spatially weighted: All available data from the following sources were used to calculate the aquifer-scale proportions—all CDPH wells in the study unit (most recent analyses that pass the quality-control tests from each well with data for that constituent during the current period, November 30, 2003, to December 1, 2006), USGS-grid wells, and understanding wells with perforation depth intervals representative of the primary aquifer system. For the spatially weighted approach, proportions were computed on a cell-by-cell basis (Isaaks and Srivastava, 1989) rather than as an average of all wells. The proportion of high relative-concentrations for each constituent for the primary aquifers was computed by (1) determining the proportion of wells with high relative-concentrations in each grid cell; and (2) averaging together the grid-cell proportions computed in step (1) (appendix C). Similar procedures were used to calculate the proportions of moderate and low relative-concentrations of constituents. The resulting proportions are spatially unbiased (Isaaks and Srivastava, 1989).

In addition, for each constituent, the detection frequencies of high and moderate relative-concentrations for individual constituents were calculated by using the same dataset as used for the spatially weighted calculations. However, these "raw" detection frequencies are not spatially unbiased because the wells in the CDPH database are not uniformly distributed throughout the USAW study unit (fig. 6). Consequently, high relative-concentrations in wells clustered in a particular area representing a small part of the primary aquifers could be given a disproportionately high weight compared to spatially unbiased methods. Raw detection frequencies are provided for reference, but were not used to assess aquifer-scale proportions (appendix C).

Aquifer-scale proportions discussed in this report primarily were estimated by using the grid-based approach, and secondarily by using the spatially weighted approach. The grid-based aquifer-scale proportions were used unless the spatially weighted proportions were significantly different. Significantly different results were defined as follows:

1. If the aquifer-scale proportion for the high category was zero using the grid-based approach, and non-zero using the spatially weighted approach, then the result from the spatially weighted approach was used. This situation can arise when the concentration of a constituent is high in a small fraction of the primary aquifers.

2. If the grid-based aquifer-scale proportion for the high category was non-zero, then the 90 percent confidence interval (based on the Jeffreys interval for the binomial distribution; Brown and others, 2001) was used to evaluate the difference. If the spatially weighted proportion was within the 90 percent confidence interval, then the grid-based proportion was used. If the spatially weighted proportion was outside the 90 percent confidence interval, then the spatially weighted proportion was used.

Aquifer-scale proportions for the moderate and low categories were determined primarily from the grid-based estimates because, for some constituents, the reporting levels for analyses in CDPH were too high to distinguish between moderate and low relative-concentrations.

Aquifer-scale proportions of high relative-concentrations also were determined for classes of constituents. The classes of organic constituents for which aquifer-scale proportions were calculated include trihalomethanes, solvents, fumigants, other VOCs, and herbicides. The classes of inorganic constituents with human-health benchmarks for which aquifer-scale proportions were calculated include trace elements, radioactive constituents, and nutrients. There are two classes of inorganic constituents with aesthetic, rather than human-health benchmarks, for which aquifer-scale proportions were calculated: salinity indicators (TDS, chloride, sulfate) and trace elements (iron, zinc, manganese, silver).

Status of Water Quality

The *status assessment* was designed to identify the constituents or classes of constituents most likely to be of water-quality concern because of their high relative-concentrations or their prevalence. USGS sample analyses, plus additional data from the CDPH database, were included in the assessment of groundwater quality for the USAW study unit. The spatially distributed, randomized approach to grid-well selection and data analysis yields a view of groundwater quality in which all areas of the primary aquifers are weighted equally; regions with a high density of groundwater use or with high density of potential contaminants were not preferentially sampled (Belitz and others, 2010).

The following discussion of the *status assessment* results is divided into three parts—inorganic, organic, and special-interest constituents. The assessment begins with a survey of how many constituents were detected at any concentration compared to the number analyzed and a graphical summary of the relative-concentrations of

constituents detected in the grid wells. Results are presented for the subset of constituents that met criteria for selection for additional evaluation based on concentration, or for organic constituents, prevalence.

The aquifer-scale proportions calculated using the spatially weighted approach were within the 90 percent confidence intervals for their respective grid-based aquifer high proportions for all 36 constituents listed in table 5, providing evidence that the grid-based and spatially weighted approaches yield statistically equivalent results. The maximum relative-concentration (sample concentration divided by the benchmark concentration) for each constituent is shown in figure 7.

Twelve inorganic constituents were detected at high relative-concentrations in one or more grid wells. These were arsenic, boron, chloride, fluoride, gross alpha radioactivity, iron, manganese, molybdenum, nitrate, total dissolved solids, uranium, and vanadium (fig. 7; table 5). Five organic constituents were detected at high relative-concentrations in one or more grid wells. These were carbon tetrachloride, 1,2-dibromo-3-chloropropane (DBCP), 1,1-dichloroethene (1,1-DCE), tetrachloroethene (PCE), and trichloroethene (TCE). Perchlorate, an inorganic special-interest constituent, also was detected at high relative-concentrations in grid wells. Constituents that were detected at moderate (but not high) relative-concentrations in at least one grid well sample included two VOCs, 1,1- and 1,2-dichloroethane, the herbicide atrazine, the major ion sulfate, and the radioactive constituents adjusted gross alpha activity, radon-222, and radium (226+228) (fig. 7).

Inorganic Constituents

Inorganic constituents generally occur naturally in groundwater, although their concentrations may be influenced by human activities (Ayotte and others, 2011). Forty-eight out of the 49 inorganic constituents analyzed by the GAMA Priority Basin Project were detected in samples from the USAW study unit; beryllium was the only inorganic constituent not detected (perchlorate is an inorganic constituent but is discussed separately as a constituent of special interest). Thirty-four of these 48 detected inorganic constituents had human-health or aesthetic benchmarks (tables 2, 6). The 14 inorganic constituents without benchmarks included four major ions (calcium, magnesium, potassium, sodium), two minor ions (bromide, iodide), three trace elements (cobalt, lithium, tungsten), two nutrient species (total nitrogen, phosphate), two radioactive constituents (gross alpha and beta 30-day counts), and silica.

Table 5. Aquifer-scale proportions from grid-based and spatially weighted approaches for constituents with high relative-concentrations during November 30, 2003–December 1, 2006, from the California Department of Public Health (CDPH) database, or with moderate or high relative-concentrations in samples collected from USGS-grid wells (November 2006–March 2007), Upper Santa Ana Watershed study unit, California GAMA Priority Basin Project.

[Grid-based aquifer proportions for organic constituents are based on samples collected by the U.S. Geological Survey from 90 grid wells during November 2006–March 2007. Spatially weighted aquifer proportions are based on CDPH data from the period December 1, 2003–November 30, 2006, in combination with grid well and understanding well data. High; concentrations greater than benchmark; moderate, concentrations less than benchmark and greater than or equal to 0.1 (for organic constituents) or 0.5 (for inorganic constituents) of benchmark; low, concentrations less than 0.1 (for organic constituents) or 0.5 (for inorganic constituents) of benchmark; MCL-US, U.S. Environmental Protection Agency maximum contaminant level; MCL-CA, CDPH maximum contaminant level; NL-CA, CDPH notification level; SMCL-CA, CDPH secondary maximum contaminant level; SMCL-US, U.S. Environmental Protection Agency secondary maximum contaminant level; mg/L, milligrams per liter; μg/L, micrograms per liter; pCi/L, picocuries per liter; THM, trihalomethane]

Constituent	Benchmark type	Benchmark value	Units	Raw detection frequency			Spatially weighted aquifer proportions			Grid-based aquifer proportions			90 percent confidence interval for grid-based high proportion [1]		Grid well detection frequency (organic and special-interest constituents only)
				Number of wells	Percent moderate (percent)	Percent high (percent)	Number of cells	Proportion moderate (percent)	Proportion high (percent)	Number of cells	Proportion moderate (percent)	Proportion high (percent)	Lower limit (percent)	Upper limit (percent)	
Trace elements and minor ions															
Aluminum	MCL-CA	1,000	μg/L	554	0.5	0.2	81	0.5	0.1	79	0.0	0.0	0.0	2.0	
Arsenic	MCL-US	10	μg/L	556	4.1	3.4	81	2.8	5.1	79	2.5	5.1	2.3	10.8	
Boron	NL-CA	1,000	μg/L	401	3.0	1.7	76	5.8	3.0	67	6.0	3.0	1.0	8.6	
Fluoride	MCL-CA	2	mg/L	563	2.8	1.2	82	3.1	2.0	79	5.1	1.3	0.3	5.5	
Lead	AL-US	15	μg/L	536	0.2	0.2	78	0.3	1.3	78	0.0	0.0	0.0	2.1	
Molybdenum	HAL-US	40	μg/L	[2]49	4.1	4.1	[2]49	4.1	4.1	[2]49	4.1	4.1	1.4	[2]11.6	
Vanadium	NL-CA	50	μg/L	338	6.2	2.4	74	8.2	2.4	62	9.7	1.6	0.4	6.9	
Trace elements - SMCL															
Iron	SMCL-CA	300	μg/L	576	2.8	3.0	82	2.7	3.7	81	3.7	2.5	0.8	7.2	
Manganese	SMCL-US	50	μg/L	576	3.1	2.8	82	4.5	4.4	81	3.7	2.5	0.8	7.2	
Radioactive constituents															
Adjusted gross alpha activity [3]	MCL-US	15[3]	pCi/L	435	2.5	0.2	74	4.0	0.3	54	3.7	0.0	0.0	3.5	
Gross alpha activity	MCL-US	15[3]	pCi/L	435	8.9	3.7	74	12	5.6	54	11.0	5.6	2.0	12	
Uranium	MCL-US	20	pCi/L	249	8.0	4.0	68	9.6	0.9	61	8.2	1.6	0.4	7.0	
Radium-226 + Radium-228	MCL-US	5	pCi/L	177	0.6	0.0	46	2.2	0.0	51	2.0	0.0	0.0	3.7	
[4]Radon-222	MCL-US	4,000	pCi/L	113	0.9	0.0	43	2.3	0.0	36	2.8	0.0	0.0	4.4	
Nutrients															
Nitrate, as nitrogen	MCL-US	10	mg/L	618	25.4	18.3	85	23.4	21.8	83	25.3	25.3	18.3	33.8	
Major ions and total dissolved solids															
Chloride	SMCL-CA	500	mg/L	571	0.9	0.4	81	3.6	1.2	81	2.5	1.2	0.3	5.3	
Sulfate	SMCL-CA	500	mg/L	571	1.4	0.0	81	1.5	0.0	81	1.2	0.0	0.0	2.0	
Total dissolved solids (TDS)	SMCL-CA	1,000	mg/L	576	13.5	2.8	81	21.9	4.7	81	24.7	4.9	2.2	10.6	

Table 5. Aquifer-scale proportions from grid-based and spatially weighted approaches for constituents with high relative-concentrations during November 30, 2003–December 1, 2006, from the California Department of Public Health (CDPH) database, or with moderate or high relative-concentrations in samples collected from USGS-grid wells (November 2006–March 2007), Upper Santa Ana Watershed study unit, California GAMA Priority Basin Project.—Continued

[Grid-based aquifer proportions for organic constituents are based on samples collected by the U.S. Geological Survey from 90 grid wells during November 2006–March 2007. Spatially weighted aquifer proportions are based on CDPH data from the period December 1, 2003–November 30, 2006, in combination with grid well and understanding well data. High; concentrations greater than benchmark; moderate, concentrations less than benchmark and greater than or equal to 0.1 (for organic constituents) of benchmark; low, concentrations less than 0.1 (for organic constituents) or 0.5 (for inorganic constituents) of benchmark; MCL-US, U.S. Environmental Protection Agency maximum contaminant level; MCL-CA, CDPH maximum contaminant level; NL-CA, CDPH notification level; SMCL-CA, CDPH secondary maximum contaminant level; SMCL-US, U.S. Environmental Protection Agency secondary maximum contaminant level; mg/L, milligrams per liter; µg/L, micrograms per liter; pCi/L, picocuries per liter; THM, trihalomethane]

Constituent	Benchmark type	Benchmark value	Units	Raw detection frequency			Spatially weighted aquifer proportions			Grid-based aquifer proportions			90 percent confidence interval for grid-based high proportion[1]		Grid well detection frequency (organic and special-interest constituents only)
				Number of wells	Percent moderate (percent)	Percent high (percent)	Number of cells	Proportion moderate (percent)	Proportion high (percent)	Number of cells	Proportion moderate (percent)	Proportion high (percent)	Lower limit (percent)	Upper limit (percent)	
Constituent of special interest															
Perchlorate	MCL-CA	6	µg/L	448	21.9	17.6	92	33.1	12.7	90	53.3	11.1	6.8	17.7	67
Solvents															
1,1-Dichloroethane	MCL-CA	5	µg/L	601	0.2	0.0	92	1.1	0.0	90	1.1	0.0	0.0	1.5	9
1,2-Dichloroethane	MCL-CA	0.5	µg/L	601	0.2	0.0	92	0.1	0.0	90	1.1	0.0	0.0	1.5	1
cis-1,2-Dichloroethene	MCL-CA	6	µg/L	601	0.3	0.0	92	0.2	0.0	90	0.0	0.0	0.0	1.5	11
Carbon tetrachloride	MCL-CA	0.5	µg/L	601	0.2	0.2	92	0.1	1.1	90	1.1	1.1	0.2	4.3	4
Tetrachloroethene (PCE)	MCL-US	5	µg/L	603	7.6	2.0	92	6.1	2.9	90	4.4	3.3	1.2	7.6	39
Trichloroethene (TCE)	MCL-US	5	µg/L	605	10.1	3.5	92	6.5	2.4	90	4.4	2.2	0.6	6.0	26
Trihalomethanes (THM)															
Chloroform	MCL-US	[5]80	µg/L	600	0.0	0.0	92	0.0	0.0	90	0.0	0.0	0.0	1.5	68
Bromodichloromethane	MCL-US	[5]80	µg/L	598	0.0	0.0	92	0.0	0.0	90	0.0	0.0	0.0	1.5	20
Fumigant															
1,2-Dibromo-3-chloropropane (DBCP)	MCL-US	0.2	µg/L	574	5.4	4.0	92	4.1	3.9	90	0.0	4.4	1.9	9.1	4
Other VOCs															
Benzene	MCL-US	1	µg/L	601	0.0	0.2	92	0.0	0.3	90	0.0	0.0	0.0	1.5	0
Dichlorodifluoromethane (CFC-12)	NL-CA	1,000	µg/L	587	0.0	0.0	92	0.0	0.0	90	0.0	0.0	0.0	1.5	10
1,1-Dichloroethene	MCL-CA	6	µg/L	602	2.7	0.8	92	3.9	0.4	90	3.3	1.1	0.2	4.3	12
Methyl tert-butyl ether (MTBE)	MCL-CA	13	µg/L	607	0.0	0.0	92	0.0	0.0	90	0.0	0.0	0.0	1.5	11
Trichlorofluoromethane (CFC-11)	MCL-CA	150	µg/L	601	0.0	0.0	92	0.0	0.0	90	0.0	0.0	0.0	1.5	11

Table 5. Aquifer-scale proportions from grid-based and spatially weighted approaches for constituents with high relative-concentrations during November 30, 2003–December 1, 2006, from the California Department of Public Health (CDPH) database, or with moderate or high relative-concentrations in samples collected from USGS-grid wells (November 2006–March 2007), Upper Santa Ana Watershed study unit, California GAMA Priority Basin Project.—Continued

[Grid-based aquifer proportions for organic constituents are based on samples collected by the U.S. Geological Survey from 90 grid wells during November 2006–March 2007. Spatially weighted aquifer proportions are based on CDPH data from the period December 1, 2003–November 30, 2006, in combination with grid well and understanding well data. High; concentrations greater than benchmark; moderate, concentrations less than benchmark and greater than or equal to 0.1 (for organic constituents) or 0.5 (for inorganic constituents) of benchmark; low, concentrations less than 0.1 (for organic constituents) or 0.5 (for inorganic constituents) of benchmark; MCL-US, U.S. Environmental Protection Agency maximum contaminant level; MCL-CA, CDPH maximum contaminant level; NL-CA, CDPH notification level; SMCL-CA, CDPH secondary maximum contaminant level; SMCL-US, U.S. Environmental Protection Agency secondary maximum contaminant level; mg/L, milligrams per liter; µg/L, micrograms per liter; pCi/L, picocuries per liter; THM, trihalomethane]

Constituent	Benchmark type	Benchmark value	Units	Raw detection frequency			Spatially weighted aquifer proportions			Grid-based aquifer proportions			90 percent confidence interval for grid-based high proportion[1]		Grid well detection frequency (organic and special-interest constituents only)
				Number of wells	Percent moderate (percent)	Percent high (percent)	Number of cells	Proportion moderate (percent)	Proportion high (percent)	Number of cells	Proportion moderate (percent)	Proportion high (percent)	Lower limit (percent)	Upper limit (percent)	
Herbicide															
Atrazine	MCL-CA	1	µg/L	522	0.4	0.2	90	0.1	0.1	87	1.1	0.0	0.0	1.5	62
Bromacil	MCL-US	70	µg/L	321	0.0	0.0	91	0.0	0.0	89	0.0	0.0	0.0	1.5	19
Diuron	HAL-US	10	µg/L	261	0.4	0.0	90	0.3	0.0	89	0.0	0.0	0.0	1.5	12
Simazine	MCL-CA	4	µg/L	552	0.0	0.0	90	0.0	0.0	87	0.0	0.0	0.0	1.5	54

[1] Based on the Jeffreys interval for the binomial distribution (Brown and others, 2001).

[2] Only GAMA data were available for molybdenum during the study period.

[3] Adjusted gross alpha activity is equal to gross alpha activity minus uranium activity. Results are presented for both adjusted and unadjusted gross alpha activity. The MCL-US applies to adjusted gross alpha activity.

[4] Proposed.

[5] The MCL-US threshold for trihalomethanes is the sum of chloroform, bromoform, bromodichloromethane, and dibromochloromethane.

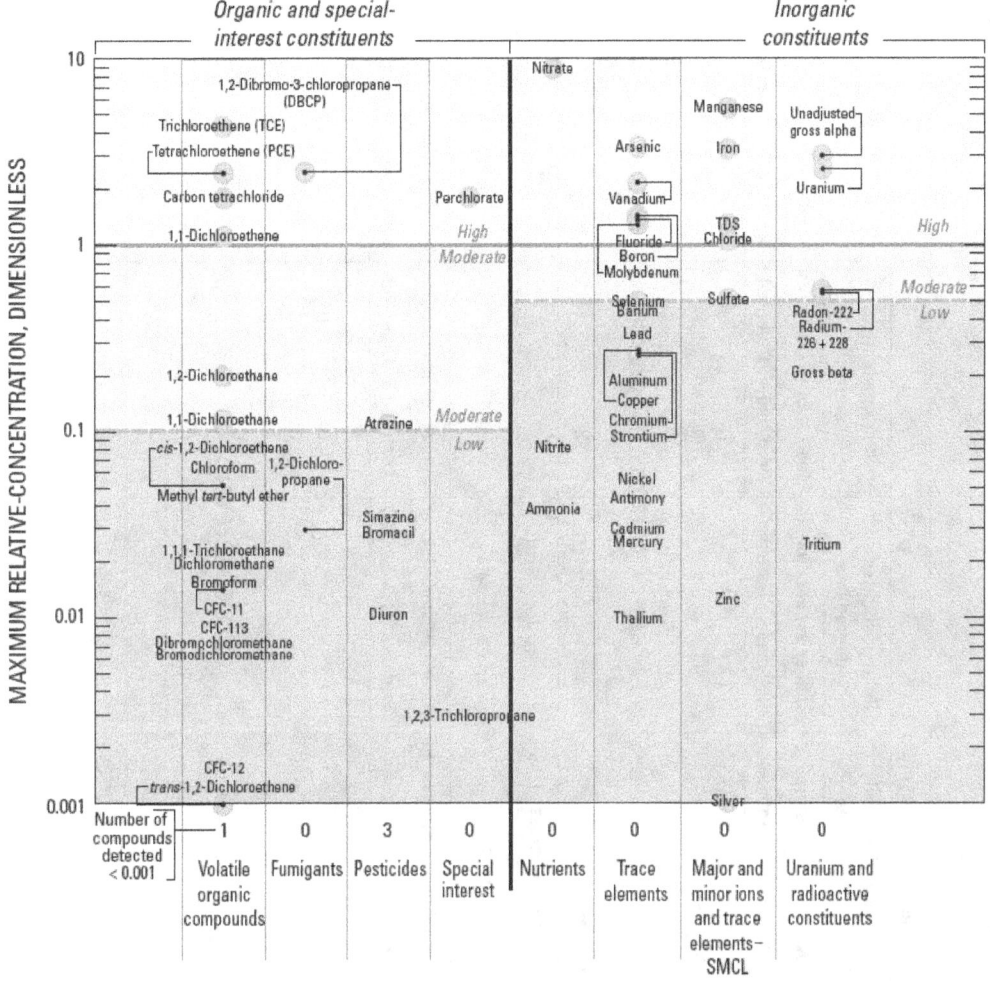

EXPLANATION

Silver   Name and center of symbol is the maximum relative-concentration for that constituent—
Unless indicated by following location line:

Abbreviations: TDS, total dissolved solids

Figure 7.   Maximum relative-concentrations of constituents detected in grid wells by constituent class, Upper Santa Ana Watershed study unit, California GAMA Priority Basin Project.

Table 6. Number of constituents analyzed and detected with associated benchmarks in each constituent class, Upper Santa Ana Watershed study unit, California GAMA Priority Basin Project, November 2006 to March 2007.

[VOC, volatile organic compound; NWQL, National Water Quality Laboratory; USEPA, U.S. Environmental Protection Agency; CDPH, California Department of Public Health; MCL, USEPA or CDPH maximum contaminant level; HAL, USEPA lifetime health advisory; NL, CDPH notification level; RSD5, USEPA risk-specific dose at 10^{-5}; AL, USEPA action level; SMCL, USEPA or CDPH secondary maximum contaminant level; TDS, total dissolved solids; TT-US, USEPA treatment technique—a required process intended to reduce the level of contamination in drinking water]

Organic constituent class

Benchmark type	Organic and special interest		Volatile organic compounds		Pesticides and degradates		Polar pesticides and degradates		Pharmaceutical compounds		Special interest	
	Analyzed	Detected	Analyzed	Detected	Analyzed	Detected	Analyzed	Detected	Analyzed	Detected	Analyzed	Detected
MCL	48	23	33	20	6	2	8	0	0	0	1	1
HAL	33	5	7	0	16	3	9	1	0	0	1	1
NL	15	1	13	1	0	0	0	0	0	0	2	0
RSD5	7	1	4	0	2	0	1	1	0	0	0	0
None	134	9	27	0	59	4	34	2	14	3	0	0
Total:	237	39	84	21	83	9	52	4	14	3	4	2

Inorganic constituent class

Benchmark type	Sum of inorganic		Major and minor ions, silica, and TDS		Nutrients		Trace elements		Radioactive		Microbial	
	Analyzed	Detected	Analyzed	Detected	Analyzed	Detected	Analyzed	Detected	Analyzed	Detected	Analyzed	Detected
MCL	21	20	1	1	2	2	11	10	7	7	1	1
HAL	3	3	0	0	1	1	2	2	0	0	0	0
NL	2	2	0	0	0	0	2	2	0	0	0	0
AL	2	2	0	0	0	0	2	2	0	0	0	0
SMCL	7	7	3	3	0	0	4	4	0	0	0	0
TT-US	0	0	0	0	0	0	0	0	0	0	3	0
None	14	14	9	9	2	2	3	3	0	0	0	0
Total:	49	48	13	13	5	5	24	23	7	7	4	1

Organic, inorganic, and microbial constituents combined [1]:	Analyzed	Detected
	290	88

[1] This total (290) differs from the total number of constituents listed on table 1 (318) because it excludes 5 field measurements, 9 dissolved gases, 7 isotopes, 6 oxidation/reduction species of metals, and dissolved organic carbon; all of which were detected, but none of which have health-based thresholds.

Thirteen inorganic constituents had high or moderate relative-concentrations in greater than 2 percent of the grid-based aquifer proportions (table 5). Eight of these thirteen inorganic constituents had high relative-concentrations in greater than 2 percent of the primary aquifers, and are discussed under their own individual headings. These are the trace elements arsenic, boron, molybdenum, iron, and manganese; the radioactive constituent gross alpha activity;

the nutrient nitrate (as nitrogen); and total dissolved solids (TDS) (table 5). Inorganic constituents with human-health benchmarks, as a group (nutrients, trace elements and minor ions, and radioactive constituents), had high aquifer proportions in 32.9 percent, moderate proportions in 29.3 percent, and low proportions (including non-detections) in 37.8 percent of the primary aquifers (table 7A).

Table 7A. Aquifer-scale proportions for inorganic constituent classes, Upper Santa Ana Watershed study unit, California GAMA Priority Basin Project.

[SMCL, secondary maximum contaminant level; values are grid based]

Constituent class	Aquifer proportion		
	Low (percent)	Moderate (percent)	High (percent)
Inorganics with human-health benchmarks			
Trace elements and minor ions	77.9	15.6	6.5
Uranium and radioactive constituents [1]	81.7	14.1	4.2
Nutrients	49.4	25.3	25.3
Any inorganic with human-health benchmarks [1]	37.8	29.3	32.9
Inorganics with aesthetic benchmarks (SMCLs)			
Salinity indicators (SMCL) [2]	75.0	20.0	5.0
Manganese and (or) iron (SMCL)	90.0	6.3	3.8
Any inorganic with an SMCL	62.5	28.8	8.8

[1] Aquifer-scale proportions for the classes uranium and radioactive constituents and any inorganic constituents with health-based benchmarks were calculated using unadjusted gross alpha activity. If adjusted gross alpha activity had been used instead, the high and moderate aquifer-scale proportions would be 1.4% and 11% respectively, for uranium and radioactive constituents, and 31% and 30%, respectively, for any inorganic constituents with health-based benchmarks.

[2] Salinity indicators consist of total dissolved solids, chloride, and sulfate.

Table 7B. Aquifer-scale proportions for organic and special-interest constituent classes, Upper Santa Ana Watershed study unit, California GAMA Priority Basin Project.

[VOCs, volatile organic compounds; values are grid based]

Constituent class	Aquifer proportion			
	Not detected	Low (percent)	Moderate (percent)	High (percent)
Organics with human-health benchmarks				
Solvents	55.6	32.2	8.9	3.3
Trihalomethanes	32.2	67.8	0.0	0.0
Fumigants	95.6	0.0	0.0	4.4
Other VOCs	71.1	24.4	3.3	1.1
Any VOC	23.3	61.1	8.9	6.7
Herbicides	31.5	67.4	1.1	0.0
Any organic with human-health benchmarks	16.7	65.6	11.1	6.7
Constituent of special interest				
Perchlorate	33.3	2.2	53.3	11.1

Trace Elements and Minor Ions

Trace elements and minor ions, as a class, had high relative-concentrations (for one or more constituents) in 6.5 percent of the primary aquifers, moderate values in 15.6 percent, and low values in 77.9 percent (table 7A). Among trace elements and minor ions, only arsenic had high relative-concentrations in greater than 5 percent of the primary aquifers (5.1 percent). Three trace elements with health-based benchmarks—arsenic, boron, and molybdenum—had high relative-concentrations (grid-based) in greater than 2 percent of the primary aquifers and are discussed under individual headings below. An additional trace element, vanadium, and the minor ion fluoride, both with health-based benchmarks, had high relative-concentrations (grid-based) in less than 2 percent of the primary aquifer (fig. 8A, table 5).

Figure 8 shows relative-concentrations in grid wells for the 14 inorganic constituents that met the criteria for discussion in this status assessment. Figure 9 is a set of maps showing the distribution and concentrations by relative-concentration category (low, medium, or high) for these constituents in USGS-grid wells and CDPH wells from November 30, 2003, to December 1, 2006.

Fluoride was detected at a high relative-concentration in 1.3 percent of the primary aquifers, and at a moderate relative-concentration in 5.1 percent (table 5; fig. 8A). High and moderate relative-concentrations were detected in a few wells of all study areas except for the Elsinore study area (fig. 9F).

Vanadium was detected at a high relative-concentration in 1.6 percent of the primary aquifers, and was detected at moderate relative-concentration in 9.7 percent (table 5; fig. 8A). Including all wells, high relative-concentrations of vanadium were detected in all study areas except for the Riverside-Arlington/Temescal study area. The Riverside-Arlington/Temescal study area did have wells with moderate relative-concentrations of vanadium (fig. 9E).

Arsenic

Arsenic had high relative-concentrations in 5.1 percent of wells and moderate values in 2.5 percent (table 5; fig. 8A). High relative-concentrations of arsenic occurred in a few wells located in all study areas except for the Yucaipa/San Timoteo study area, where relative-concentrations were moderate and low (fig. 9A).

Boron

Boron was detected at high relative-concentrations in 3.0 percent of the primary aquifers, and at moderate relative-concentrations in 6.0 percent of the primary aquifers (table 5; fig. 8A). High and moderate relative-concentrations of boron were detected in the southwestern areas of the Cucamonga/Chino and Riverside-Arlington/Temescal study areas, and in the central part of the San Jacinto study area (fig. 9B).

Molybdenum

Molybdenum data were limited to the 49 GAMA grid wells that were sampled for trace elements because CDPH does not collect data for molybdenum. High relative-concentrations of molybdenum were detected in 2 of the 49 GAMA grid wells (4.1 percent of the primary aquifers), with that same proportion (4.1 percent) of moderate relative-concentrations of molybdenum in the primary aquifers (table 5; fig. 8A). High and moderate relative-concentrations were measured in the Elsinore study area (fig. 9C). A moderate relative-concentration of molybdenum also was detected in the San Jacinto study area.

Uranium and Radioactive Constituents

Concentrations of uranium and radioactive constituents generally are low in the USAW study unit with a few exceptions. Radium (combined 226 and 228) and radon-222 had moderate relative-concentrations in 2.0 and 2.8 percent of the primary aquifers, respectively (table 5; fig. 8B). The single moderate relative-concentration of radium was detected in the San Jacinto study area in a CDPH-grid well (fig. 9H). A moderate relative-concentration of radon-222 was detected in the Elsinore study area in a USGS GAMA-grid well (fig. 9I). Relative-concentrations for radon-222 were calculated by using the higher of two proposed MCLs for this constituent—4,000 picocuries per liter (pCi/L), a level which assumes that the State or local water agency has an approved multimedia mitigation program to address radon in indoor air (U.S. Environmental Protection Agency, 1999). If the determination of relative-concentrations had been based on the lower proposed MCL of 300 pCi/L (applicable in the absence of such a program), 58 percent of USAW grid wells that were sampled for radon-222 would have had high relative-concentrations. However, the number of cells with radon-222 data is small. Therefore, the calculated aquifer proportions may not be representative of the study unit for this constituent.

Uranium was detected at a high relative-concentration in 1.6 percent of the primary aquifers, and at moderate relative-concentrations in 8.2 percent of the primary aquifers (table 5; fig. 8B). High relative concentrations of uranium were detected in the Bunker Hill/Cajon/Rialto-Colton study area. Moderate relative-concentrations of uranium were found in all study areas except for the San Jacinto and Elsinore study areas and were concentrated in the Riverside-Arlington/Temescal study area (fig. 9D).

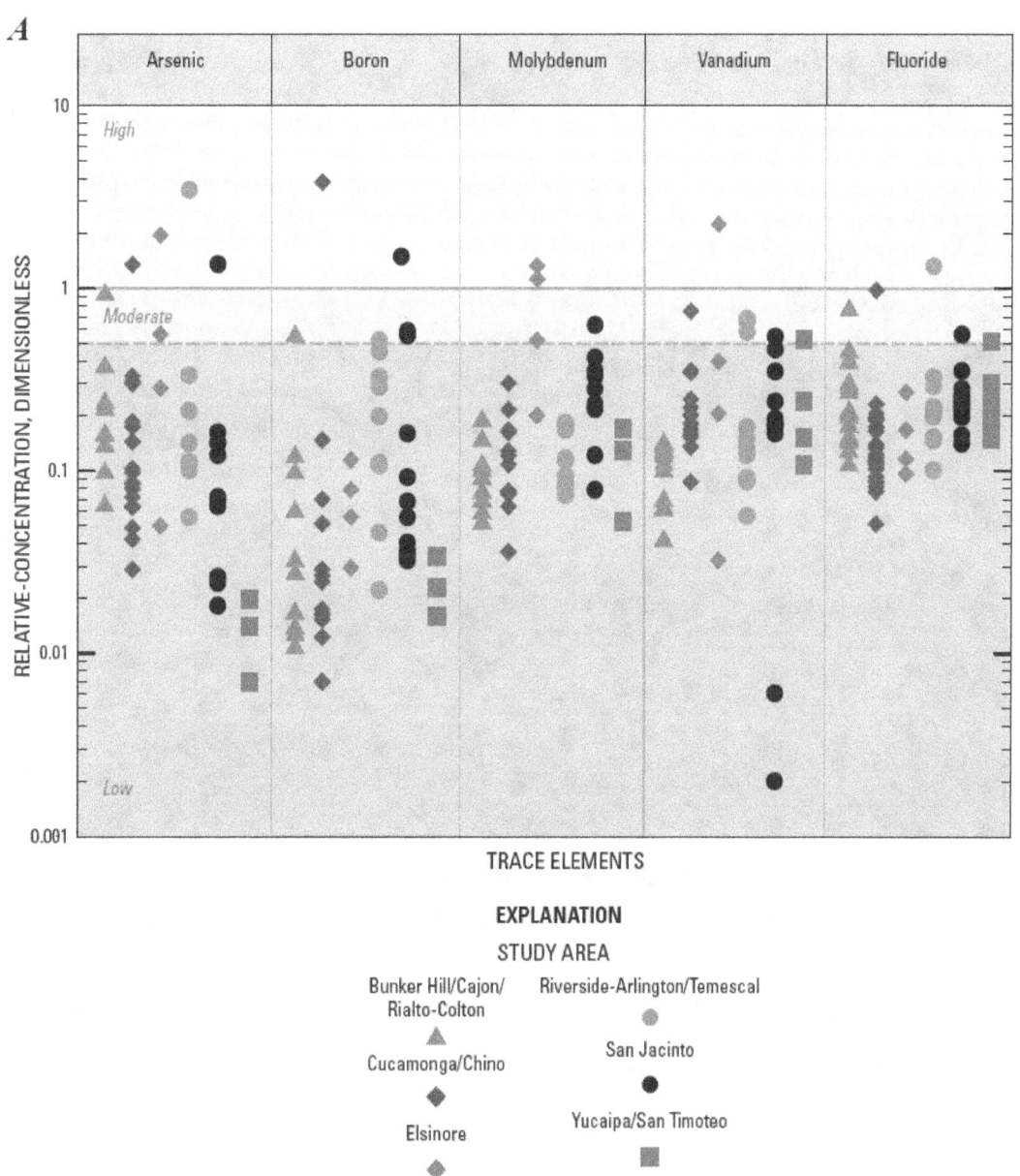

Figure 8. Relative-concentrations of (*A*) selected trace elements, (*B*) radioactive constituents, (*C*) nutrients, and (*D*) major and minor ions in grid wells (USGS and CDPH), Upper Santa Ana Watershed study unit, GAMA Priority Basin Project.

B

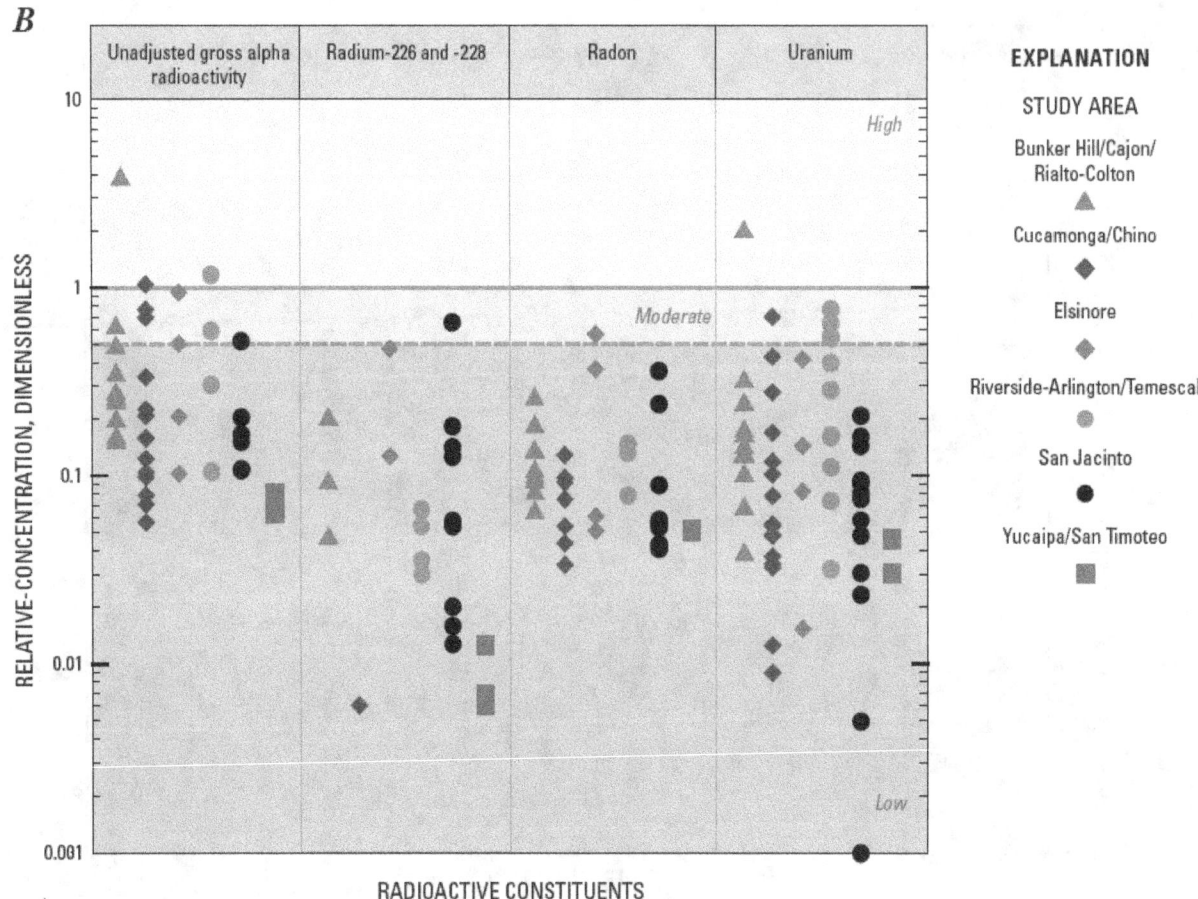

Figure 8.—Continued.

C

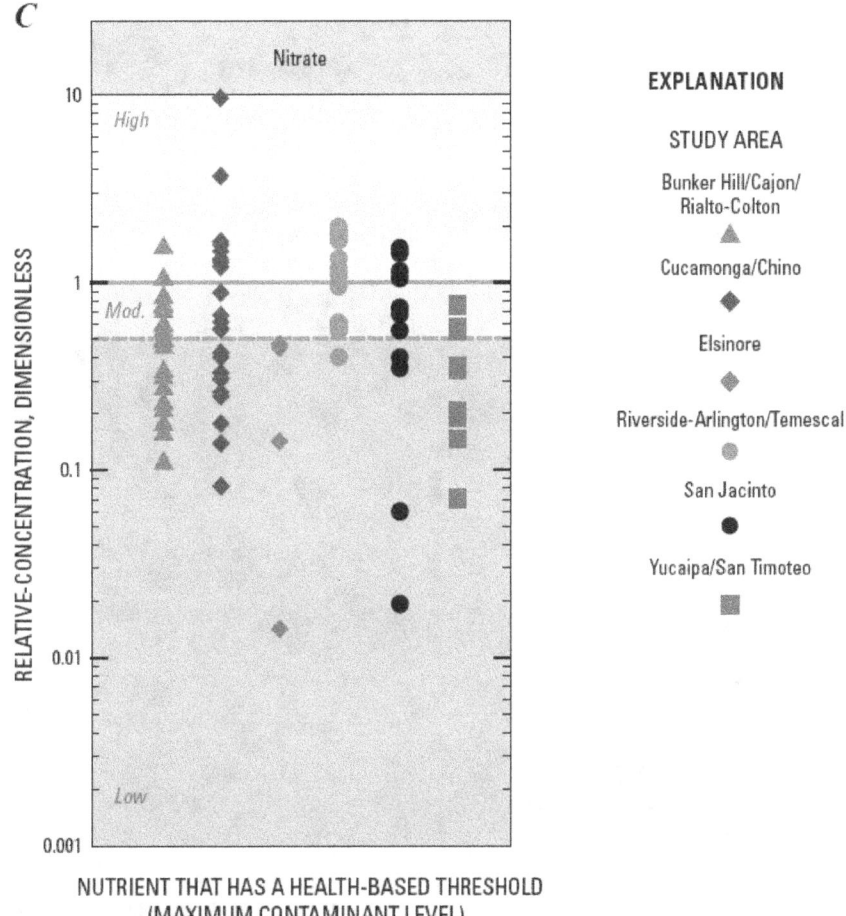

NUTRIENT THAT HAS A HEALTH-BASED THRESHOLD
(MAXIMUM CONTAMINANT LEVEL)

Figure 8.—Continued.

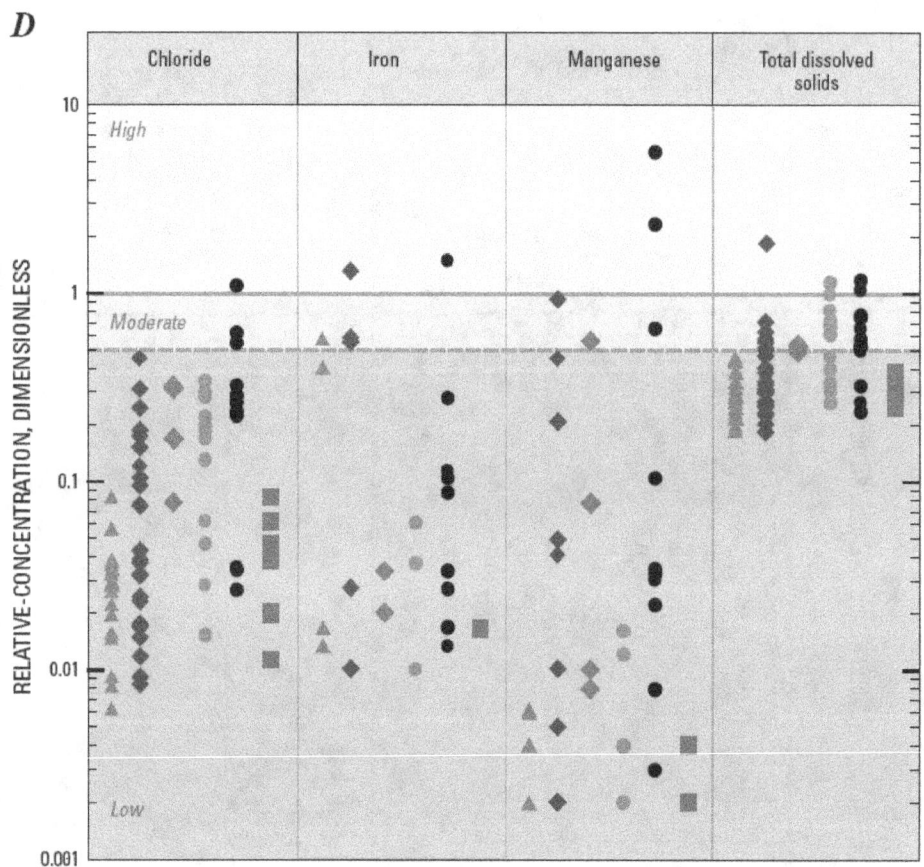

MAJOR AND MINOR IONS THAT HAVE SECONDARY MAXIMUM CONTAMINANT LEVELS

EXPLANATION

STUDY AREA

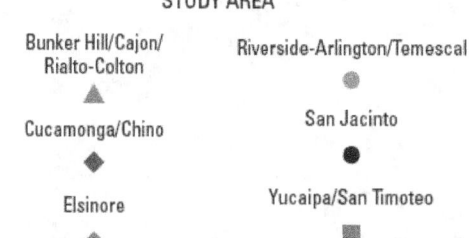

Figure 8.—Continued.

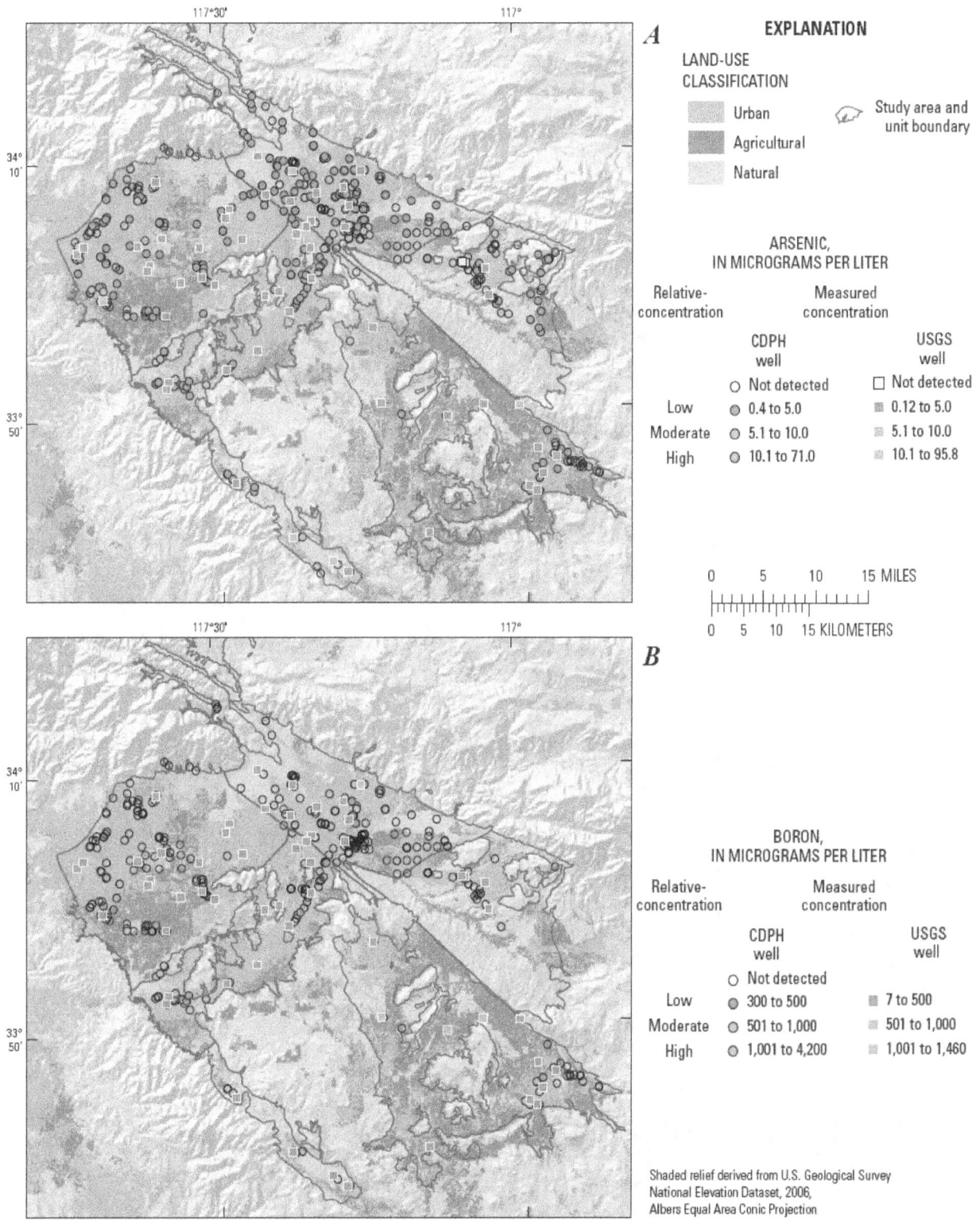

Figure 9. Relative-concentrations for selected inorganic constituents, (*A*) arsenic, (*B*) boron, (*C*) molybdenum, (*D*) uranium, (*E*) vanadium, (*F*) fluoride, (*G*) gross alpha radioactivity, (*H*) radium, (*I*) radon, (*J*) nitrate, (*K*) chloride, (*L*) iron, (*M*) manganese, and (*N*) total dissolved solids (TDS), Upper Santa Ana Watershed study unit, GAMA Priority Basin Project, USGS grid and understanding wells representative of the primary aquifer and the most recent analysis during November 30, 2003, to December 1, 2006, for CDPH wells.

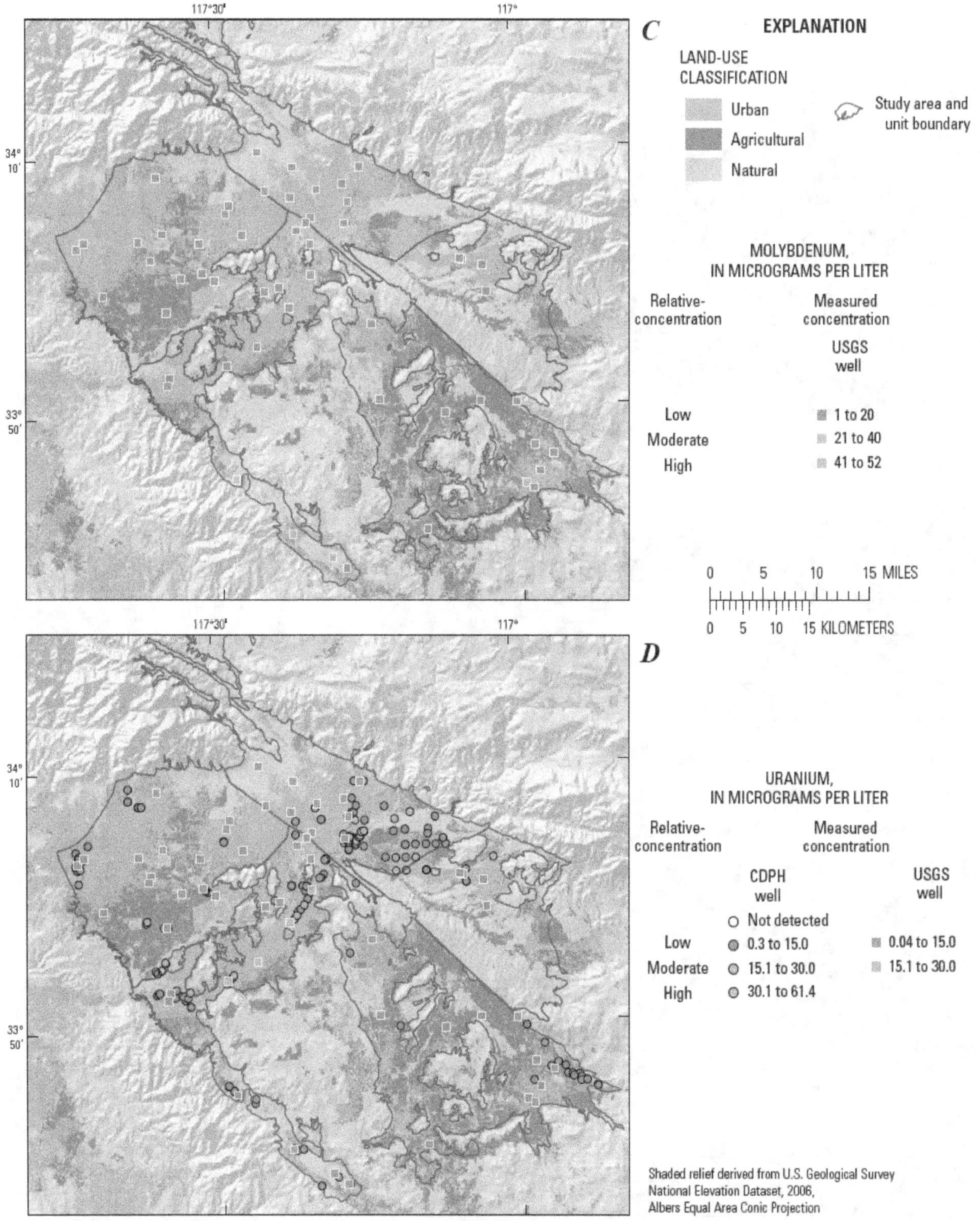

EXPLANATION

C

LAND-USE
CLASSIFICATION

Urban

Agricultural

Natural

Study area and
unit boundary

MOLYBDENUM,
IN MICROGRAMS PER LITER

Relative-concentration	Measured concentration
	USGS well
Low	1 to 20
Moderate	21 to 40
High	41 to 52

```
0        5        10      15 MILES
|‖‖‖‖‖‖‖‖|‖‖‖‖‖‖‖‖|‖‖‖‖‖‖‖‖|
0    5        10     15 KILOMETERS
```

D

URANIUM,
IN MICROGRAMS PER LITER

Relative-concentration	Measured concentration	
	CDPH well	USGS well
	○ Not detected	
Low	◉ 0.3 to 15.0	0.04 to 15.0
Moderate	◉ 15.1 to 30.0	15.1 to 30.0
High	◉ 30.1 to 61.4	

Shaded relief derived from U.S. Geological Survey
National Elevation Dataset, 2006,
Albers Equal Area Conic Projection

Figure 9.—Continued.

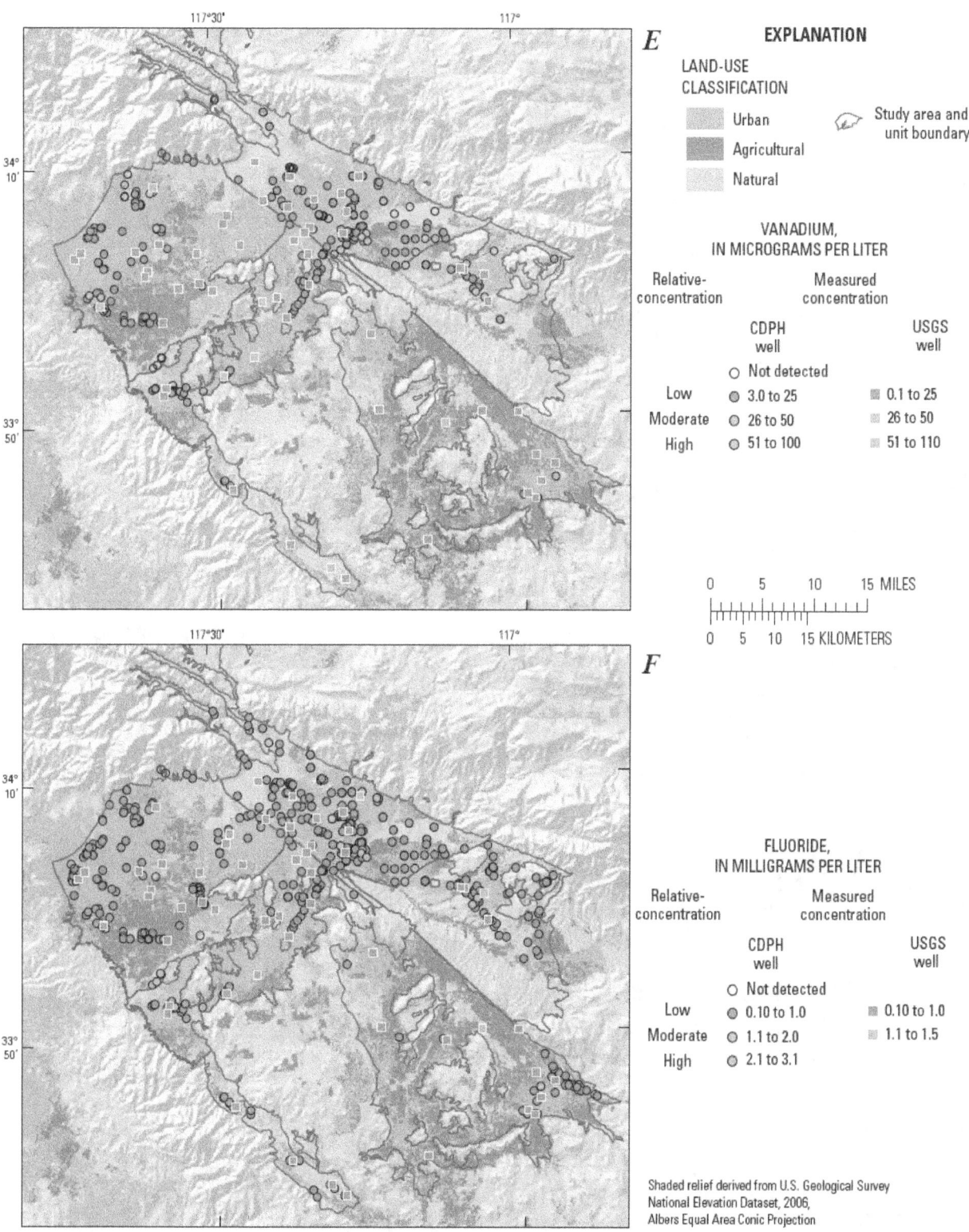

EXPLANATION

LAND-USE CLASSIFICATION

Urban

Agricultural

Natural

Study area and unit boundary

VANADIUM, IN MICROGRAMS PER LITER

Relative-concentration	Measured concentration	
	CDPH well	USGS well
	○ Not detected	
Low	◉ 3.0 to 25	▦ 0.1 to 25
Moderate	◉ 26 to 50	▦ 26 to 50
High	◉ 51 to 100	▦ 51 to 110

0   5   10   15 MILES

0   5   10   15 KILOMETERS

FLUORIDE, IN MILLIGRAMS PER LITER

Relative-concentration	Measured concentration	
	CDPH well	USGS well
	○ Not detected	
Low	◉ 0.10 to 1.0	▦ 0.10 to 1.0
Moderate	◉ 1.1 to 2.0	▦ 1.1 to 1.5
High	◉ 2.1 to 3.1	

Shaded relief derived from U.S. Geological Survey National Elevation Dataset, 2006, Albers Equal Area Conic Projection

Figure 9.—Continued.

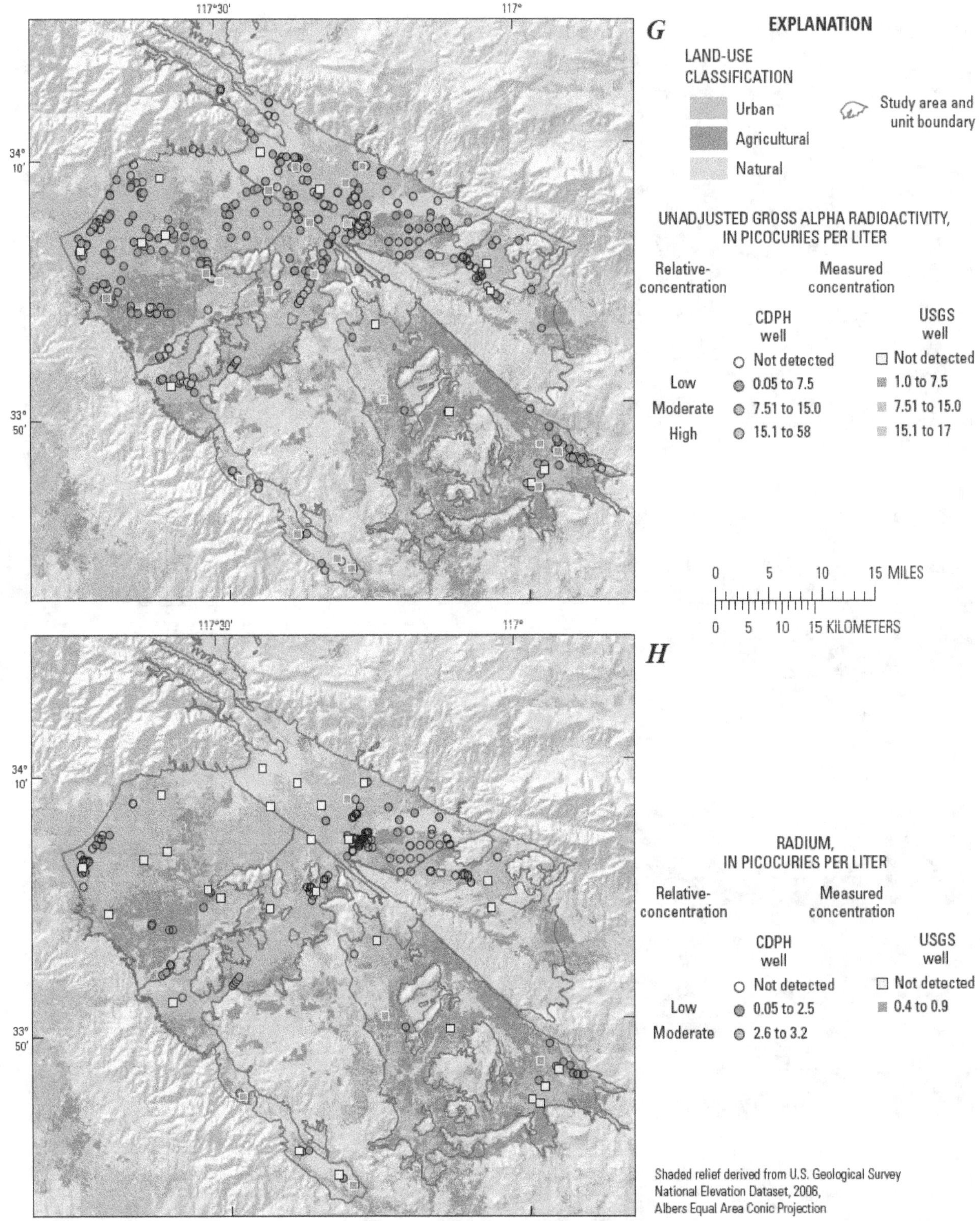

G

EXPLANATION

LAND-USE CLASSIFICATION

- Urban
- Agricultural
- Natural

Study area and unit boundary

UNADJUSTED GROSS ALPHA RADIOACTIVITY, IN PICOCURIES PER LITER

Relative-concentration	Measured concentration	
	CDPH well	**USGS well**
	○ Not detected	☐ Not detected
Low	◉ 0.05 to 7.5	▨ 1.0 to 7.5
Moderate	◉ 7.51 to 15.0	▨ 7.51 to 15.0
High	◉ 15.1 to 58	▨ 15.1 to 17

```
0        5        10       15 MILES
|-+-+-+-+-|-+-+-+-+-|-+-+-+-+-|
0        5        10       15 KILOMETERS
```

H

RADIUM, IN PICOCURIES PER LITER

Relative-concentration	Measured concentration	
	CDPH well	**USGS well**
	○ Not detected	☐ Not detected
Low	◉ 0.05 to 2.5	▨ 0.4 to 0.9
Moderate	◉ 2.6 to 3.2	

Shaded relief derived from U.S. Geological Survey
National Elevation Dataset, 2006,
Albers Equal Area Conic Projection

Figure 9.—Continued.

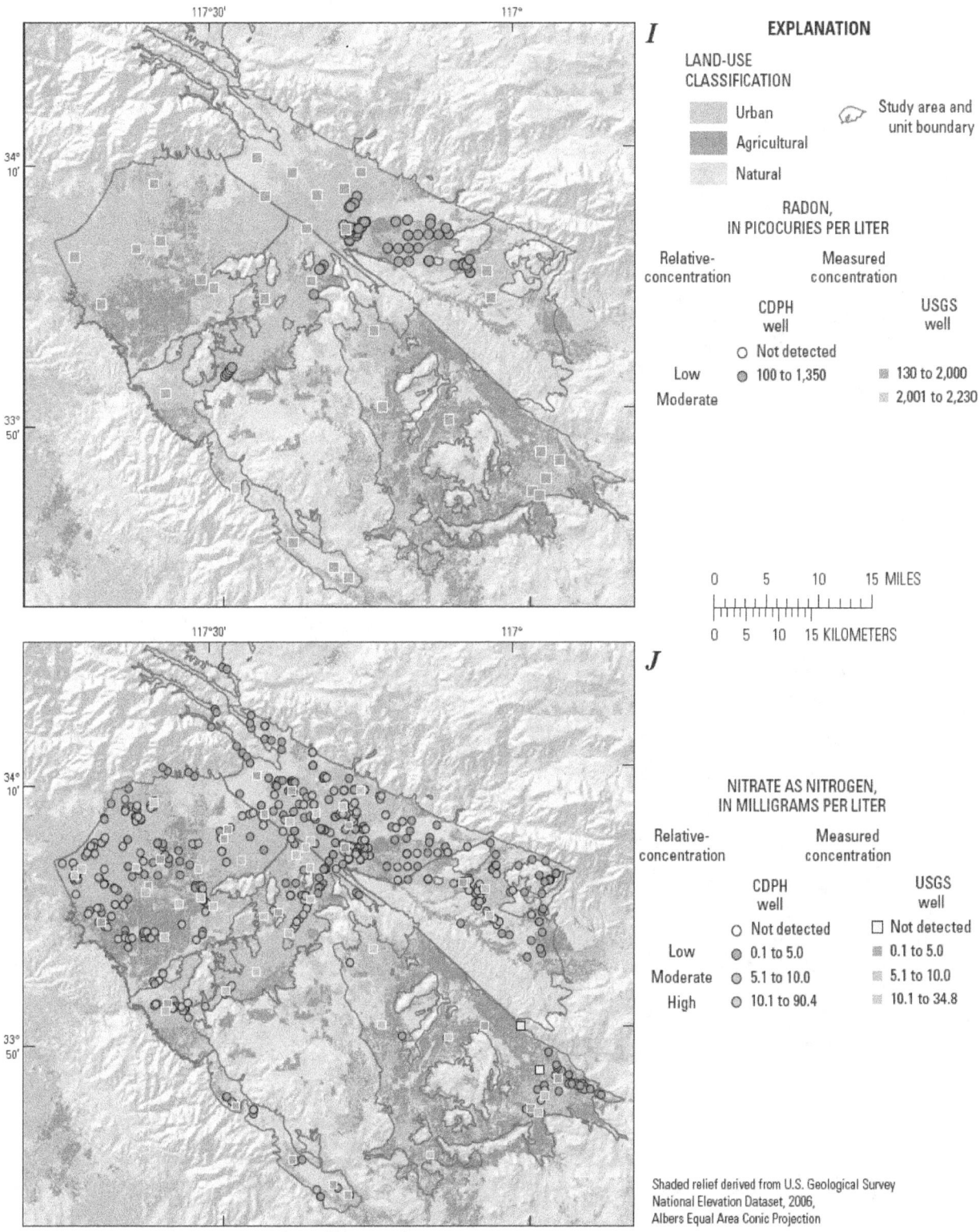

EXPLANATION

LAND-USE
CLASSIFICATION

Urban

Agricultural

Natural

Study area and
unit boundary

RADON,
IN PICOCURIES PER LITER

Relative-concentration	Measured concentration	
	CDPH well	USGS well
	○ Not detected	
Low	◉ 100 to 1,350	▨ 130 to 2,000
Moderate		▨ 2,001 to 2,230

0   5   10   15 MILES

0   5   10   15 KILOMETERS

NITRATE AS NITROGEN,
IN MILLIGRAMS PER LITER

Relative-concentration	Measured concentration	
	CDPH well	USGS well
	○ Not detected	☐ Not detected
Low	◉ 0.1 to 5.0	▨ 0.1 to 5.0
Moderate	◉ 5.1 to 10.0	▨ 5.1 to 10.0
High	◉ 10.1 to 90.4	▨ 10.1 to 34.8

Shaded relief derived from U.S. Geological Survey
National Elevation Dataset, 2006,
Albers Equal Area Conic Projection

Figure 9.—Continued.

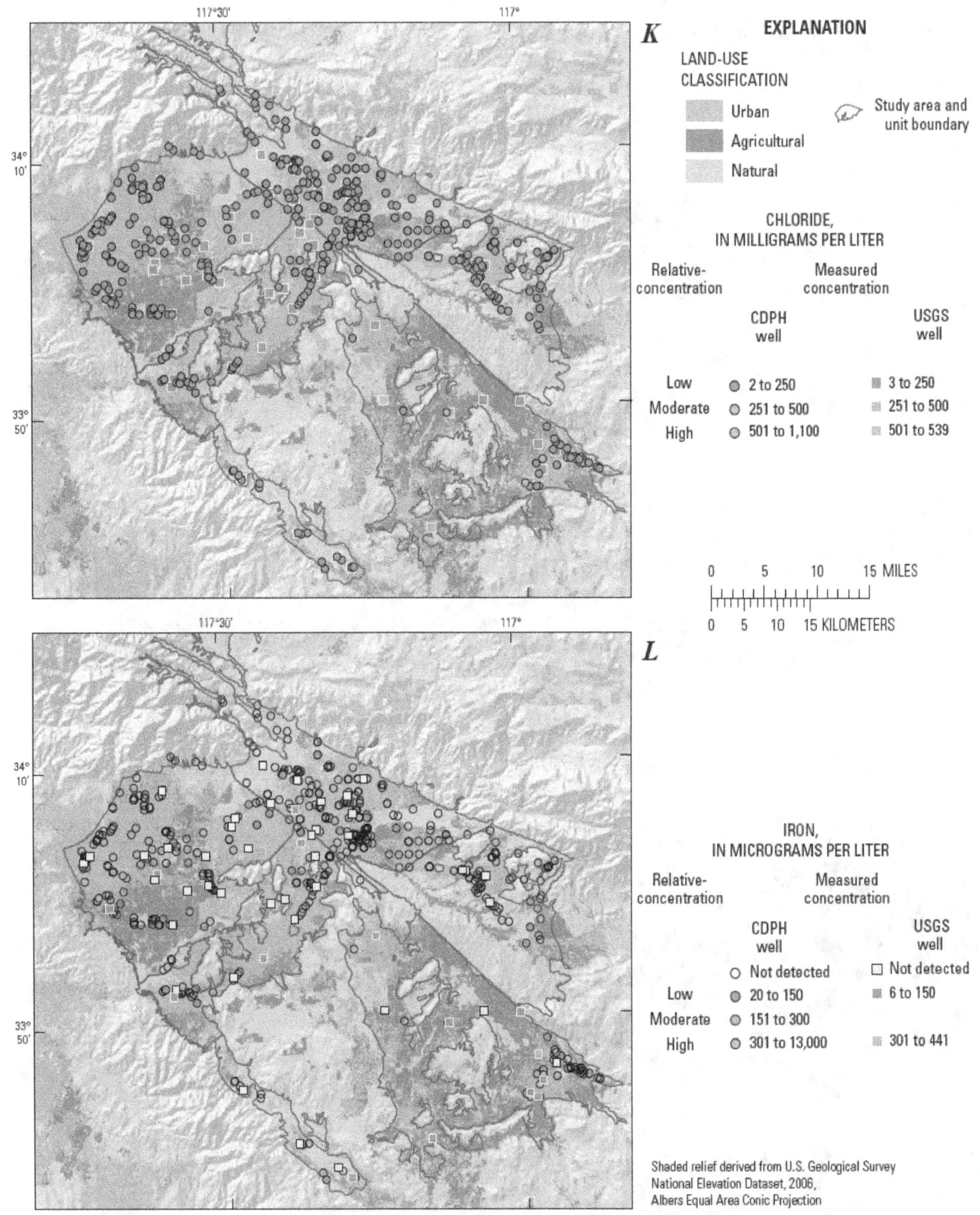

K

EXPLANATION

LAND-USE CLASSIFICATION

Urban

Agricultural

Natural

Study area and unit boundary

CHLORIDE, IN MILLIGRAMS PER LITER

Relative-concentration	Measured concentration	
	CDPH well	USGS well
Low	2 to 250	3 to 250
Moderate	251 to 500	251 to 500
High	501 to 1,100	501 to 539

0   5   10   15 MILES

0   5   10   15 KILOMETERS

L

IRON, IN MICROGRAMS PER LITER

Relative-concentration	Measured concentration	
	CDPH well	USGS well
	Not detected	Not detected
Low	20 to 150	6 to 150
Moderate	151 to 300	
High	301 to 13,000	301 to 441

Shaded relief derived from U.S. Geological Survey
National Elevation Dataset, 2006,
Albers Equal Area Conic Projection

Figure 9.—Continued.

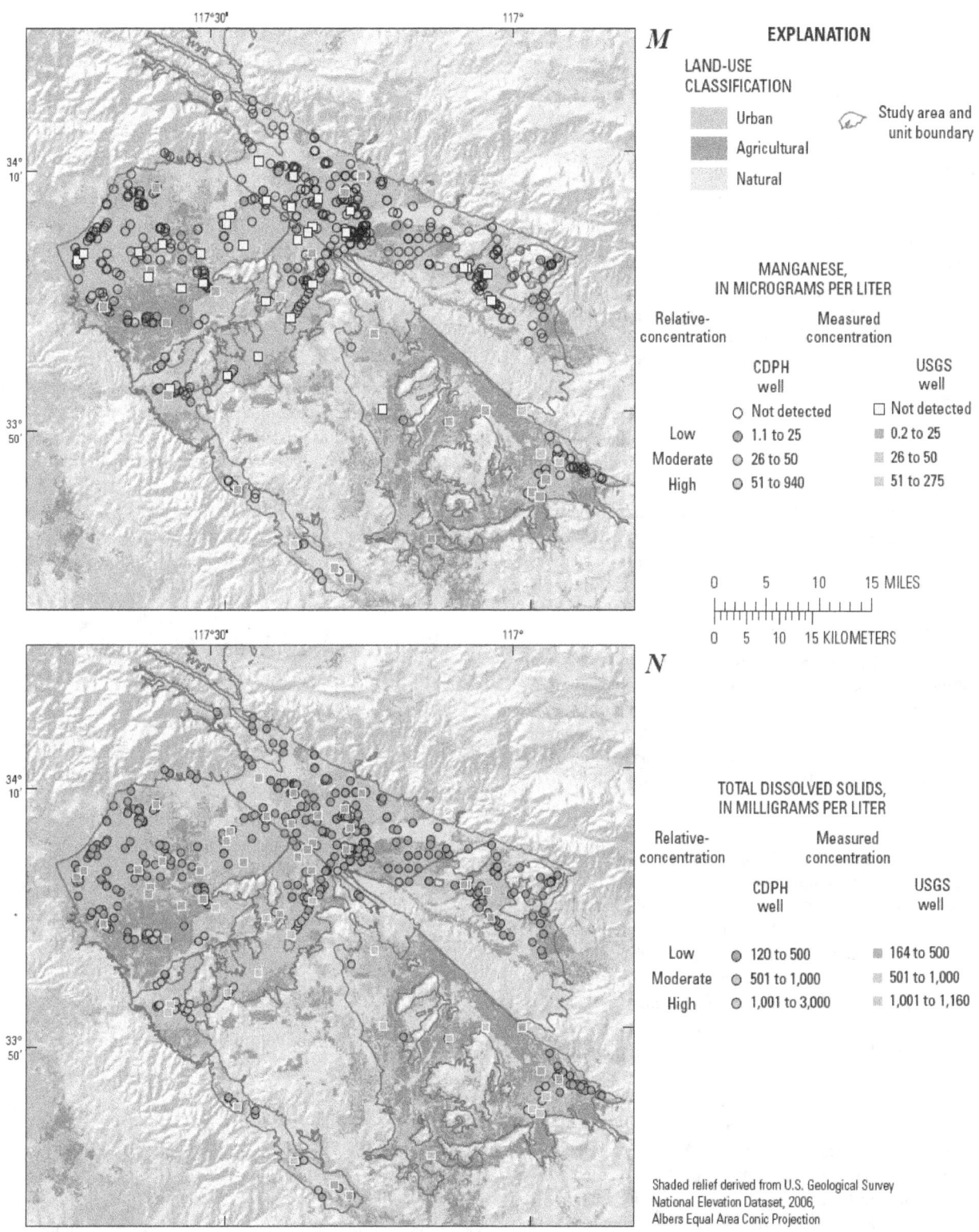

EXPLANATION

LAND-USE
CLASSIFICATION

Urban

Agricultural

Natural

Study area and
unit boundary

MANGANESE,
IN MICROGRAMS PER LITER

Relative-concentration	Measured concentration	
	CDPH well	USGS well
	○ Not detected	☐ Not detected
Low	◉ 1.1 to 25	▨ 0.2 to 25
Moderate	◉ 26 to 50	▨ 26 to 50
High	◉ 51 to 940	▨ 51 to 275

TOTAL DISSOLVED SOLIDS,
IN MILLIGRAMS PER LITER

Relative-concentration	Measured concentration	
	CDPH well	USGS well
Low	◉ 120 to 500	▨ 164 to 500
Moderate	◉ 501 to 1,000	▨ 501 to 1,000
High	◉ 1,001 to 3,000	▨ 1,001 to 1,160

Shaded relief derived from U.S. Geological Survey
National Elevation Dataset, 2006,
Albers Equal Area Conic Projection

Figure 9.—Continued.

Gross Alpha Activity

Gross alpha activity had high relative-concentrations in 5.6 percent of the primary aquifers, and moderate relative-concentrations in 11 percent of the primary aquifers. These results are for unadjusted gross alpha activity. The MCL-US (15 pCi/L) for gross alpha activity applies to adjusted gross alpha activity, which is equal to the measured gross alpha activity minus uranium activity (U.S. Environmental Protection Agency, 2000). Data collected by USGS-GAMA and data compiled in the CDPH database are reported as gross alpha activity without correction for uranium activity. Gross alpha is used a screening tool to determine whether additional radioactive constituents must be analyzed (California Department of Public Health, 2012). For regulatory purposes, analysis of uranium is only required if gross alpha activity is greater than 15 pCi/L; therefore, the CDPH database contains more data for gross alpha activity than for uranium. As a result, it is not always possible to calculate adjusted gross alpha activity. For this reason, gross alpha data without correction for uranium are the primary data used in the status assessments made by USGS-GAMA for the Priority Basin Project. Examination of data from samples having USGS-GAMA data for uranium and gross alpha indicated that, in the absence of data for uranium, uncorrected gross alpha data likely provide a more accurate estimate of the aquifer-scale proportions for uranium and radioactive constituents as a class, than do adjusted gross alpha data (Miranda Fram, USGS California Water Science Center, written commun., 2012). All of the groundwater with high or moderate relative-concentrations of gross alpha activity also had high or moderate relative-concentrations of uranium.

USGS-GAMA reports data for gross alpha activity counted 72 hours and 30 days after sample collection. Regulatory sampling for gross alpha activity permits use of quarterly composite samples (U.S. Environmental Protection Agency, 2000; California Department of Public Health, 2012); thus, the USGS-GAMA gross alpha 30-day count data may be more appropriate to use when combining USGS-GAMA and CDPH datasets. Gross alpha activity in a groundwater sample may change with time after sample collection due to radioactive decay and ingrowth (activity may increase or decrease depending on sample composition and holding time) (Arndt, 2010).

Most data for uranium in the CDPH database are reported as activities in units of picocuries per liter, and the majority of uranium data gathered by USGS-GAMA are reported as concentrations in units of micrograms per liter. The factor used to convert uranium mass concentration to uranium activity depends on the isotopic composition of the uranium (U.S. Environmental Protection Agency, 2000). This report uses a conversion factor of 0.79.

Nutrients

Three of the four species of nitrogen analyzed for in this study have health-based thresholds—ammonia, nitrite, and nitrate. One of these, nitrite plus nitrate (hereinafter referred to as nitrate), was frequently detected at concentrations above its MCL. All detections of the other nitrogen species with health-based thresholds—nitrite and ammonia—were at low relative-concentrations. The only phosphorus species measured, orthophosphate, has no drinking-water aesthetic or regulatory benchmark.

Nitrate

Nitrate was detected at high relative-concentrations in 25.3 percent of the primary aquifers, with this same proportion of primary aquifers having moderate relative-concentrations (table 5; fig. 8C). Thus, just over half of USAW grid wells had high or moderate relative-concentrations of nitrate. High relative-concentrations of nitrate were detected in all USAW study areas except for the Elsinore study area (fig. 9J). High and moderate relative-concentrations were most prevalent in the Cucamonga/Chino and Riverside-Arlington/Temescal study areas.

Constituents with Secondary Maximum Contaminant Level Benchmarks

CDPH has established non-enforceable benchmarks, secondary maximum contaminant levels (SMCL-CA), which are based on aesthetic properties rather than on human-health concerns for selected constituents. For total dissolved solids (TDS) and the major ions chloride and sulfate, CDPH defines a "recommended" and an "upper" SMCL-CA. The "upper" SMCL-CA benchmarks were used for computing relative-concentrations for this report. CDPH defines a single benchmark for iron and a single benchmark for manganese.

Chloride was detected at a high relative-concentration in one grid well (1.2 percent of the primary aquifers), and was detected at moderate relative-concentrations in 2.5 percent of the primary aquifers (table 5; fig. 8D). High and moderate relative-concentrations of chloride were detected only in the San Jacinto study area (fig. 9K).

Total Dissolved Solids

Relative-concentrations of total dissolved solids (TDS) were high in 4.9 percent of the primary aquifers, and moderate in 24.7 percent (table 5; fig. 8D). High and moderate relative-concentrations of TDS were most prevalent in the San Jacinto, Riverside-Arlington/Temescal, and Cucamonga/Chino study areas. Relative-concentrations of TDS in the Bunker Hill/Cajon/Rialto-Colton, Yucaipa/San Timoteo, and Elsinore study areas were mostly low (fig. 9N).

Iron and Manganese

Two trace elements with SMCL-CAs, iron and manganese, were each detected at high relative-concentrations in 2.5 percent of the primary aquifers, and at moderate relative-concentrations in 3.7 percent (table 5; fig. 8D). High or moderate relative-concentrations of iron and manganese were observed in all of the study areas with one exception. The Yucaipa/San Timoteo study area had no grid wells with high relative-concentrations of iron and no grid wells with high or moderate relative-concentrations of manganese (figs. 8D, 9L, and 9M).

Organic Constituents

In this report, organic compounds are organized by constituent class, including four classes of volatile organic compounds (VOCs), two classes of pesticides, and pharmaceuticals. VOCs may be present in paints, solvents, fuels, refrigerants, or disinfected water, and are characterized by their tendency to evaporate. VOCs are classified here as trihalomethanes, solvents, fumigants, or other VOCs (including gasoline additives and refrigerants). Pesticides are used to control weeds (herbicides), insects (insecticides), or fungi (fungicides) in agricultural, urban, and suburban settings. Pesticides are classified here as herbicides or insecticides and fungicides. Once released into the environment, pesticides are transformed, over time, by a variety of chemical, photochemical, and biologically mediated reactions into other compounds, which are referred to in this report as degradates (Gilliom and others, 2006).

In contrast to the nearly ubiquitous inorganic constituents, only 39 of the 237 organic and special-interest constituents analyzed were detected at concentrations greater than their respective long-term method detection limits (dissolved organic carbon is not included in this discussion). These compounds included 21 VOCs, 13 pesticides and pesticide degradates, 3 pharmaceuticals, and 2 special-interest constituents (perchlorate and 1,2,3-trichloropropane). Most of the organic and special-interest constituents detected (30 of the 39) have human-health benchmarks (table 6). Of the nine detected organic compounds lacking benchmarks, three are herbicide degradates, and three are pharmaceuticals; the other three are the fungicide, metalaxyl, and the herbicides, norflurazon and pendimethalin. Two of the three herbicide degradates that were detected, 2-chloro-4-isopropylamino-6-amino-s-triazine (deethylatrazine) and 2-chloro-6-ethylamino-4-amino-s-triazine (deisopropyl-atrazine), are degradates of atrazine and other triazine herbicides (Gilliom and others, 2006). Atrazine and another triazine herbicide, simazine, were each frequently detected (detected in at least 10 percent of samples), and both have health-based benchmarks. The third herbicide degradate detected, 3,4-dichloroaniline, is a degradate of diuron, which was frequently detected and has a health-based benchmark. Diuron is among the most heavily used herbicides in the Santa Ana Basin (Kent and others, 2005). Three pharmaceutical compounds were detected at concentrations greater than or equal to method detection limits in one or more samples from the USAW study unit. Fram and Belitz (2011) present all results for pharmaceutical compounds in groundwater samples collected for 28 of the GAMA Priority Basin Project study units (studies that took place during May 2004 through March 2010). The three pharmaceuticals that were detected in USAW groundwater samples were acetaminophen (analgesic), carbamazepine (antiepileptic), and caffeine.

An additional 13 pesticide compounds and 2 VOCs were reported as detections by the NWQL in USAW samples at concentrations less than their long-term method detection limits (LT-MDL) (Kent and Belitz, 2009). For purposes of this study, these compounds are not considered detections because detections with concentrations less than the LT-MDL have greater than a 1 percent probability of being falsely-positive detections (Childress and others, 1999).

Among all organic constituents with human-health benchmarks, the proportion of USAW primary aquifers with high concentrations was 6.7 percent (table 7B), which reflects the high relative-concentrations of the solvents PCE (3.3 percent) and TCE (2.2 percent), and the fumigant 1,2-dibromo-3-chloropropane (DBCP) (4.4 percent) (table 5). The proportion of the aquifers having moderate relative-concentrations of organic constituents with human-health benchmarks was 11.1 percent (table 7B).

Organic constituents were detected in 83 percent of the primary aquifers. Fifteen organic compounds had maximum relative-concentrations that were high or moderate in more than 2 percent of the primary aquifers, detection frequencies greater than 10 percent, or both. Five organic compounds—the solvents carbon tetrachloride, PCE, and TCE, the fumigant DBCP, and 1,1-dichloroethene (1,1-DCE, organic synthesis)—were detected at maximum concentrations greater than their health-based thresholds (maximum relative-concentrations greater than 1, or "high") (figs. 7, 10). Three of these five—1,1-DCE, TCE, and PCE—also had detection frequencies greater than or equal to 10 percent (table 5; fig. 10). Ten additional organic compounds with health-based thresholds were detected in at least 10 percent of the primary aquifers, although all but one (atrazine) were detected only at low maximum relative-concentrations. These 10 compounds were the THMs chloroform and bromodichloromethane, the solvent cis-1,2-dichloroethene (cis-1,2-DCE), the gasoline oxygenate methyl tert-butyl ether (MTBE), the refrigerants trichlorofluoromethane (CFC-11) and dichlorodifluoromethane (CFC-12), and the herbicides atrazine, bromacil, diuron, and simazine (fig. 11). Atrazine was detected at a moderate relative-concentration in one grid well.

The detection frequency of one or more VOCs in the 90 grid wells was 77 percent. Of the 21 VOCs detected, 14 were detected only at low relative-concentrations (fig. 10). Nine VOCs had a detection frequency of at least 10 percent in the grid wells: chloroform, PCE, TCE, bromodichloromethane, 1,1-DCE, cis-1,2-DCE, MTBE, CFC-11, and CFC-12 (figs. 10, 11; table 5).

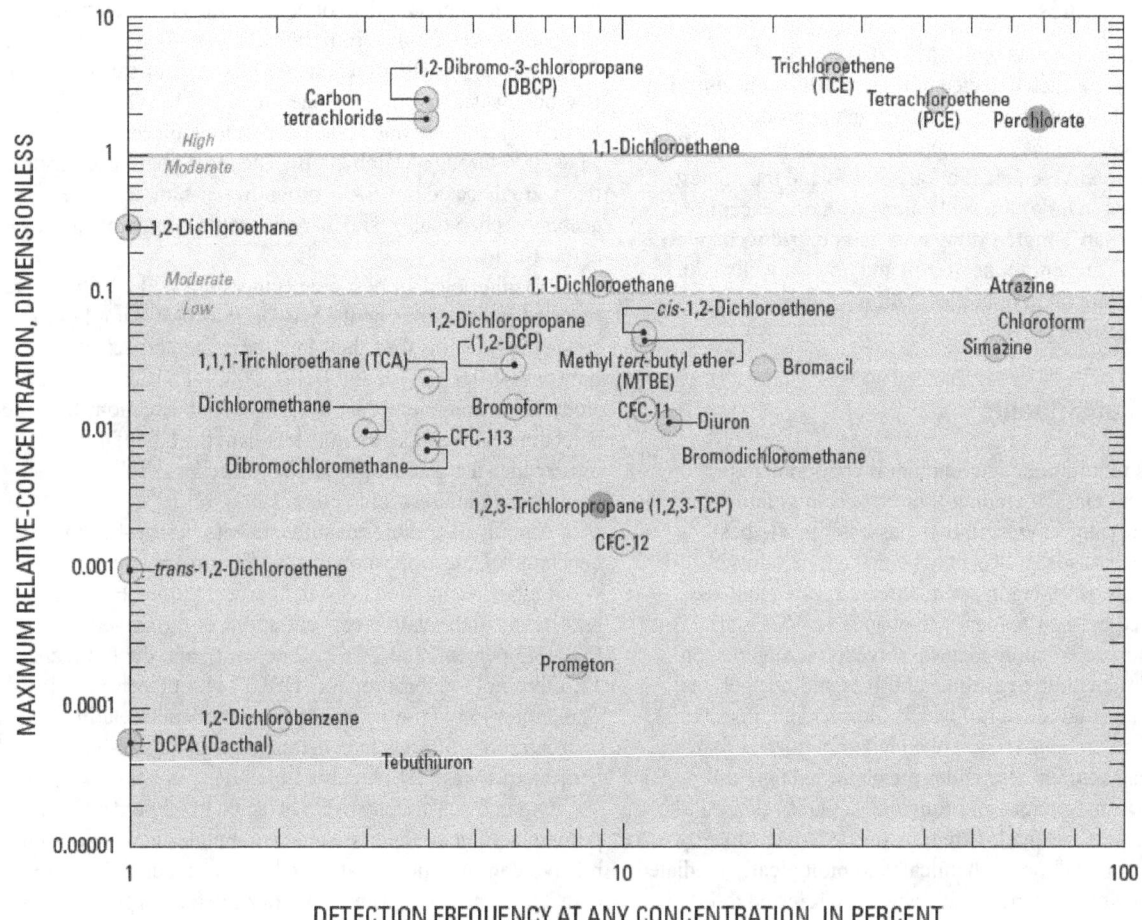
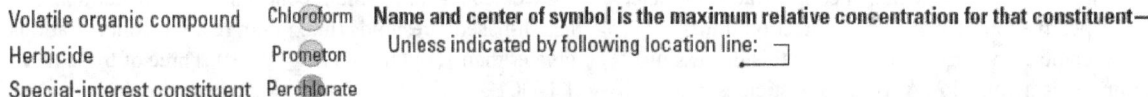

EXPLANATION

Volatile organic compound   Chloroform   **Name and center of symbol is the maximum relative concentration for that constituent—**
Herbicide   Prometon   Unless indicated by following location line:
Special-interest constituent   Perchlorate

Figure 10. Detection frequency and maximum relative-concentration of organic and special-interest constituents detected in USGS-grid wells in the Upper Santa Ana Watershed study unit, California GAMA Priority Basin Project.

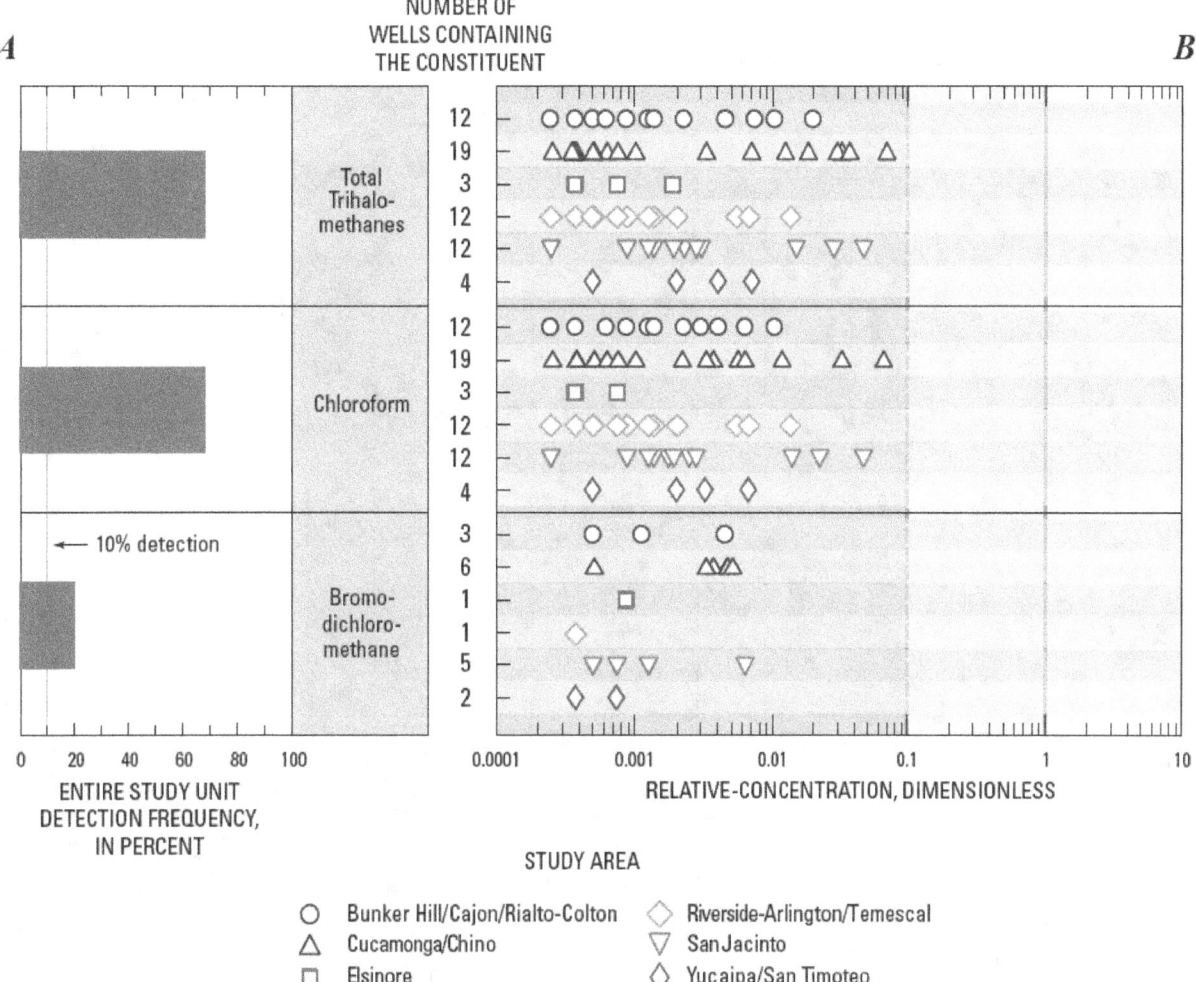

Figure 11 (*A–L*).   Detection frequency (bar charts) and relative-concentrations (dot plots) of selected organic and special-interest constituents in grid wells in the Upper Santa Ana Watershed study unit, California GAMA Priority Basin Project, November 2006–March 2007.

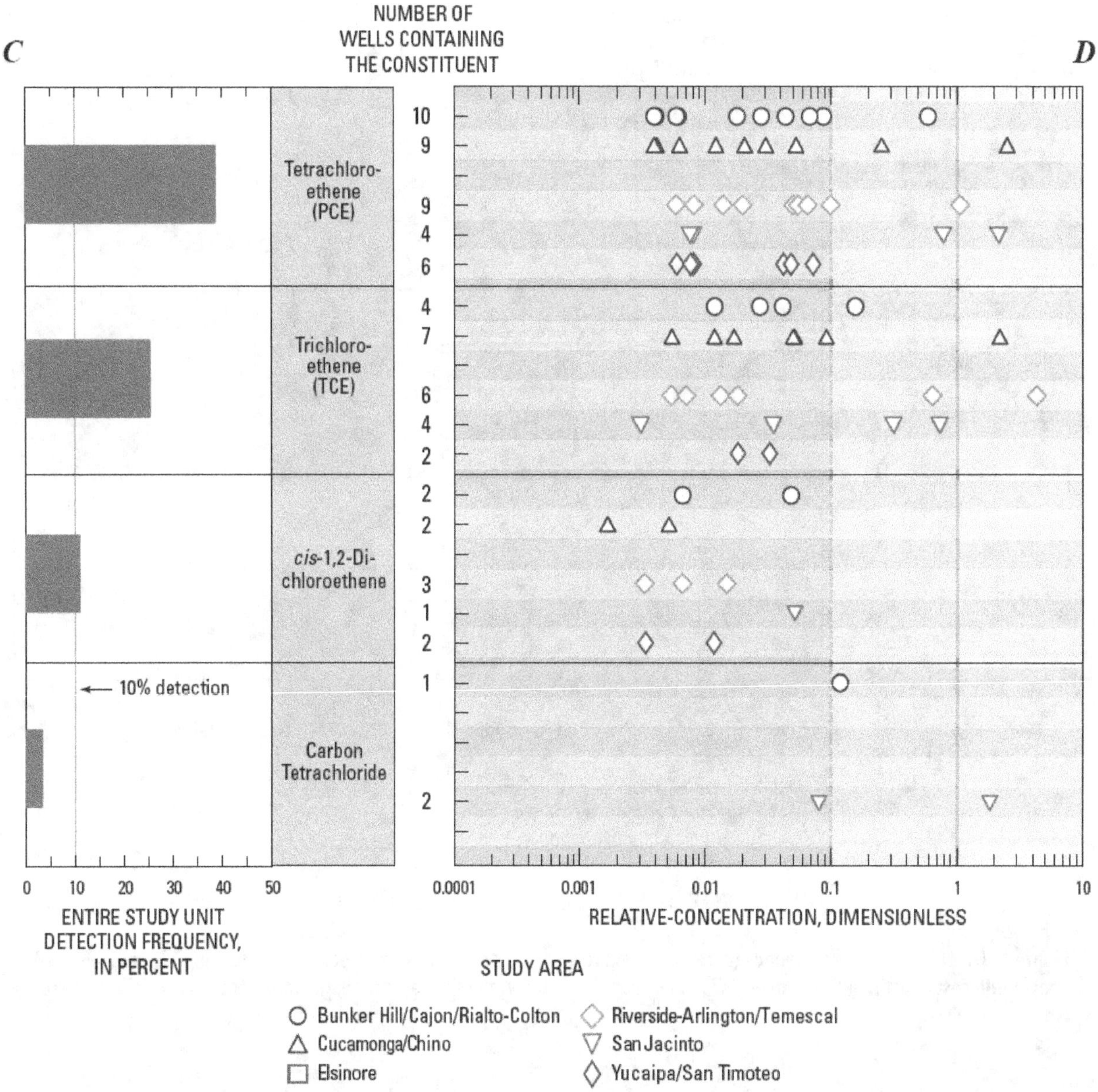

Figure 11.—Continued.

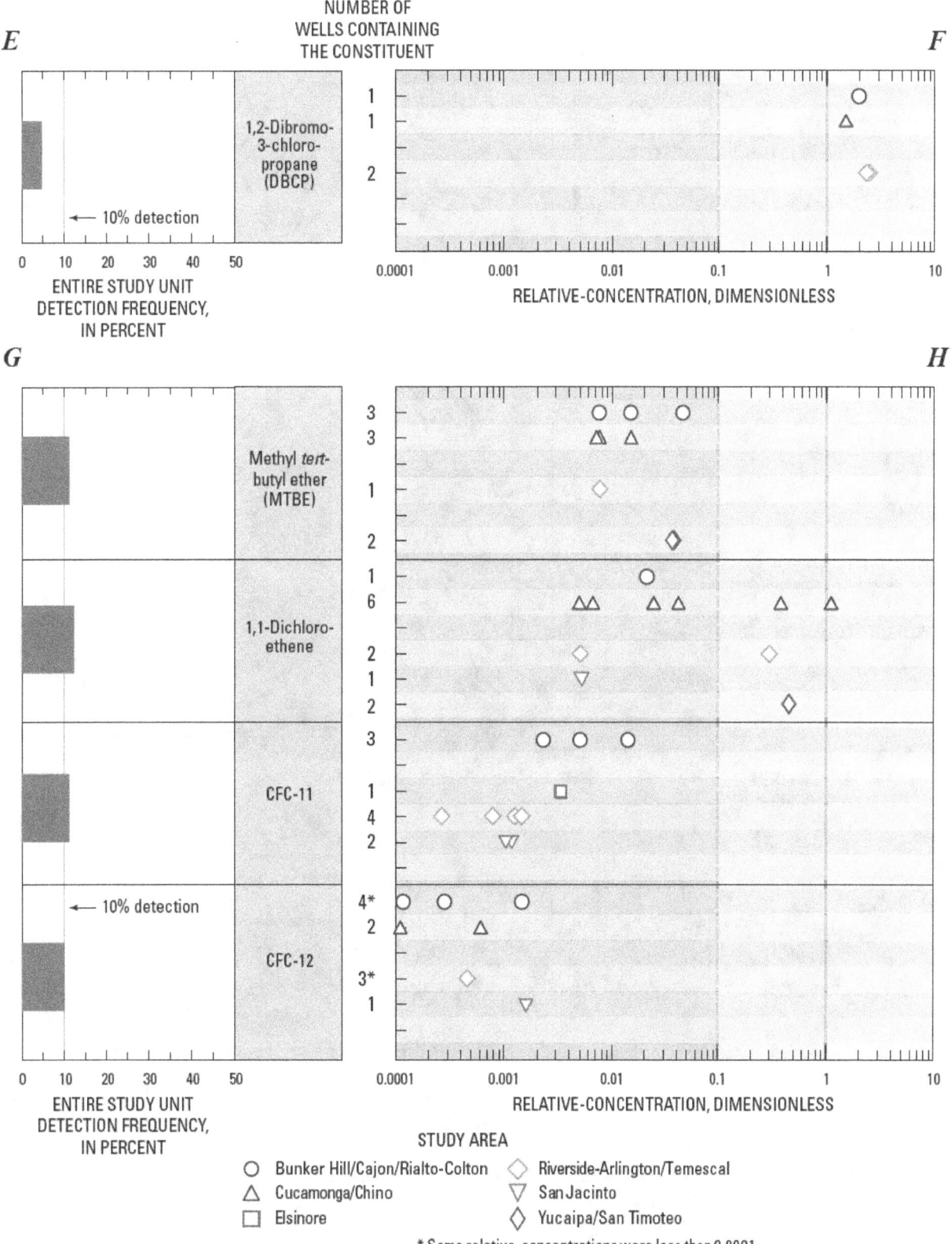

Figure 11.—Continued.

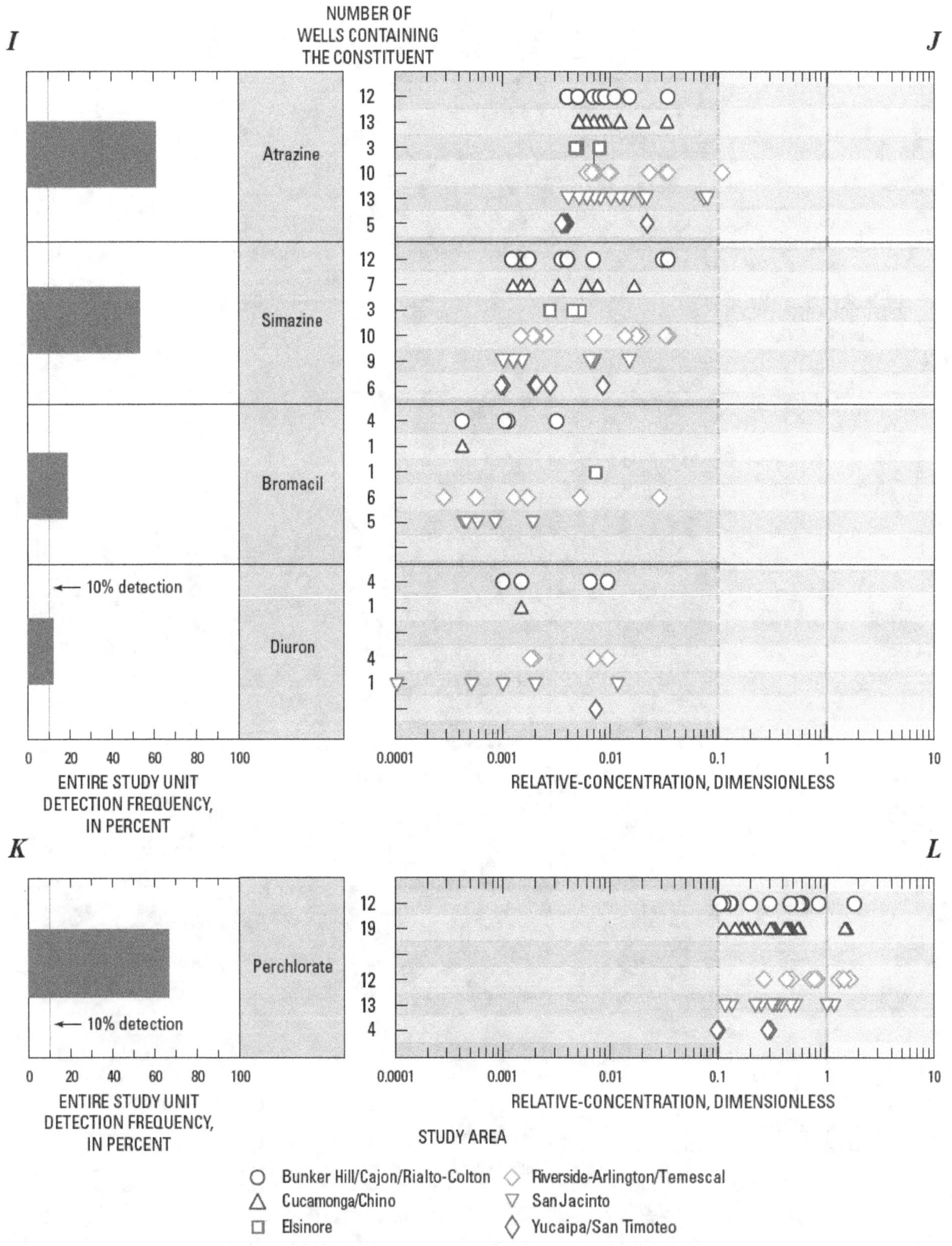

Figure 11.—Continued.

One or more pesticides or pesticide degradates were detected in 68 percent of the primary aquifers. Of the 135 pesticides and pesticide degradates analyzed, 13 were detected. Seven of the detected pesticides or pesticide degradates were parent compounds with benchmarks, three were parent compounds without a benchmark, and three were degradates without benchmarks. All concentrations of pesticides with human-health benchmarks were below the benchmarks. One pesticide, atrazine, was detected at a moderate relative-concentration in one sample (fig. 11J). All of the other pesticide compounds were detected only at low relative-concentrations. Four herbicides—atrazine, simazine, bromacil, and diuron—and two herbicide degradates—2-chloro-4-isopropylamino-6-amino-s-triazine (deethylatrazine) and 1,4-dichloroaniline—were detected in more than 10 percent of the primary aquifers. The four herbicide parent compounds are discussed in further detail below. Discussion of the frequently occurring herbicide degradates is beyond the scope of this study. The individual constituents that were not detected and the wells sampled in the USAW study unit are listed in Kent and Belitz (2009).

Trihalomethanes

Water used for drinking water and other household uses in domestic and municipal systems commonly is disinfected with some form of chlorine disinfectant. As a side effect to disinfecting the water, the chlorine reacts with organic matter to produce trihalomethanes (THMs) and other chlorinated and/or brominated disinfection byproducts. Potential urban sources of THMs include recharge from landscape irrigation with disinfected water, leakage from distribution or sewer systems, and industrial and commercial sources (Ivahnenko and Barbash, 2004). The four chlorinated and/or brominated THMs (chloroform, bromodichloromethane, dibromochloromethane, and bromoform) were detected at low relative-concentrations. Two of these—chloroform and bromodichloromethane—were frequently detected (detected in at least 10 percent of samples).

Chloroform was detected in 68 percent of primary aquifers (figs. 10, 11A). Chloroform was detected throughout the USAW study unit (fig. 12A). The study area with the lowest detection frequency was Yucaipa/San Timoteo at 44 percent (Kent and Belitz, 2009). In contrast, chloroform was detected in all (100 percent) of the grid wells in the Riverside-Arlington/Temescal study area. In general, the analytical methods used by the CDPH to analyze for organic compounds have higher detection limits than those used by the NWQL; therefore, non-detections are more common in CDPH wells for chloroform, as well as for other organic compounds (fig. 12).

Bromodichloromethane was another THM frequently detected at low relative-concentrations in USAW grid wells (figs. 10, 11). Bromodichloromethane was detected throughout the USAW study unit, with a detection frequency of about 20 percent in the study unit (Kent and Belitz, 2009). However, in contrast to the 100 percent detection frequency for chloroform in the Riverside-Arlington/Temescal study area, bromodichloromethane was detected in only 1 of the 12 grid wells (8 percent) in the Riverside-Arlington/Temescal study area (fig. 12B).

The health-based threshold for chlorinated/brominated THMs is applied to the summed concentration of the four compounds; sometimes referred to as "total trihalomethanes" for regulatory purposes. Similar to concentrations of the individual THMs, concentrations of the summed compounds were low during the current study period. Chloroform was the predominant compound making up total THMs and was present in all samples where a THM was detected. One well in the CDPH database for the USAW study unit had a historically high relative-concentration for total THMs in a sample collected before December 2003 (table 4).

Solvents

Solvents are a class of VOCs used for a variety of industrial, commercial, and domestic purposes. Of the 29 solvents analyzed for in this study, 10 were detected (table 3A of Kent and Belitz, 2009). Four of these met selection criteria for further discussion here, either because they were frequently detected (detected in more than 10 percent of the grid wells), or because they were detected at a high or moderate relative-concentrations in more than 2 percent of the primary aquifers. The solvent compounds that were frequently detected in USAW grid wells were PCE, TCE, and cis-1,2-dichloroethene. The solvent carbon tetrachloride, though not frequently detected, had high and moderate grid-based aquifer proportions of 1.1 percent each (table 5). Solvents, as a group, had a high aquifer-scale proportion of 3.3 percent, and a moderate aquifer-scale proportion of 8.9 percent (table 7B).

Carbon tetrachloride (tetrachloromethane) was detected in three USAW grid wells, with high, moderate, and low relative-concentrations each occurring in one well. The moderate relative-concentration occurred in a well in the Bunker Hill/Cajon/Rialto-Colton study area, and the high and low relative-concentrations occurred in the San Jacinto study area (figs. 11, 12E). Cis-1,2-dichloroethene was detected at low relative-concentrations in 11 percent of USAW grid wells, and was detected in all of the USAW study areas except for Elsinore (figs. 11, 12F). This solvent, frequently detected in USAW grid wells, had no high values during the current study period, but did have a historically high value in the CDPH database from the period prior to November 30, 2003 (table 4). Other solvents with no high values during the current period of study, but with historically high values in the CDPH database, were 1,1-dichloroethane, dichloromethane (methylene chloride), and 1,1,2-trichloroethane (table 4).

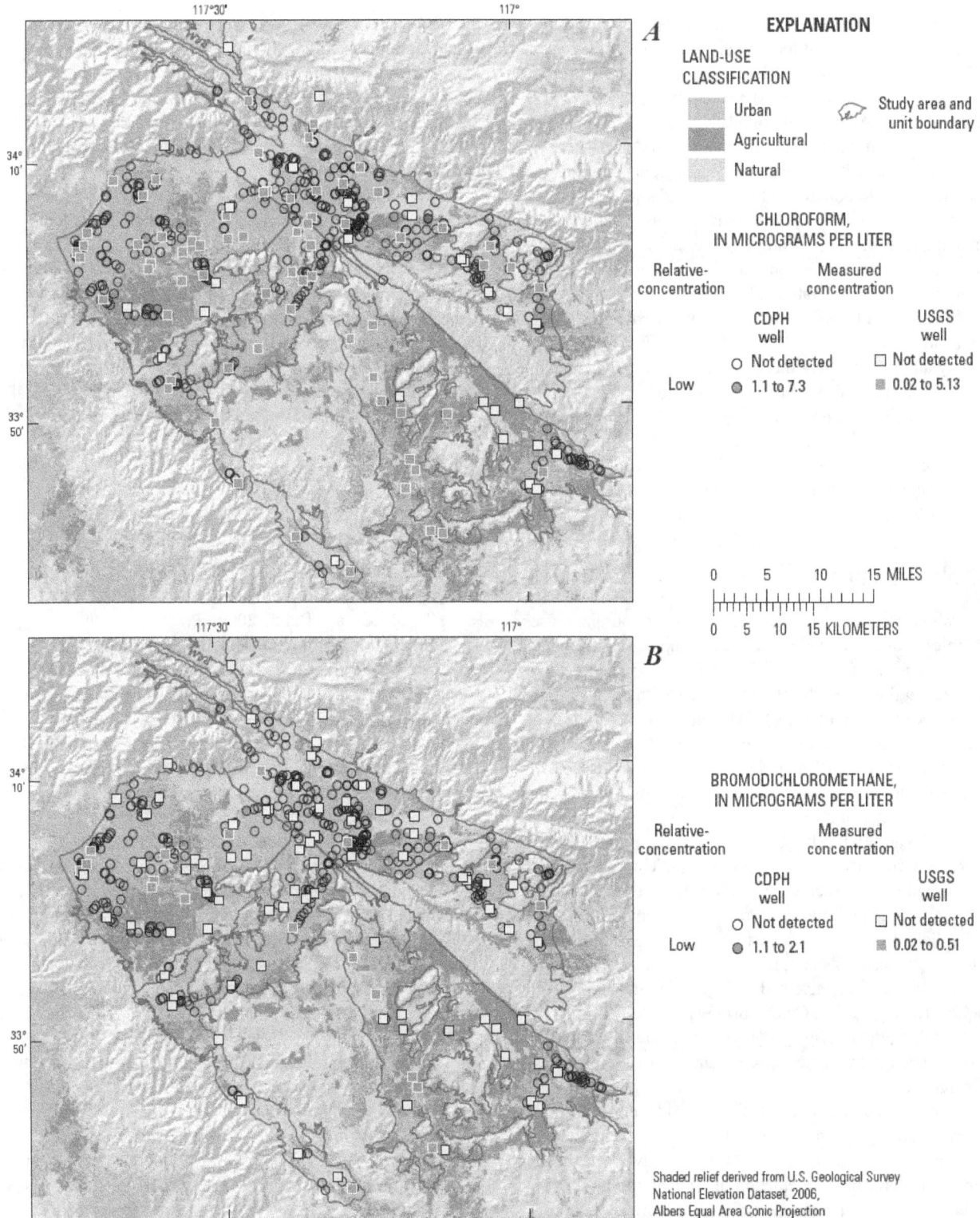

Figure 12 (A–P).   Relative-concentrations for selected organic and special-interest constituents, (A) chloroform, (B) bromodichloromethane, (C) tetrachloroethene, (D) trichloroethene, (E) carbon tetrachloride, (F) cis-1,2-dichloroethene, (G) 1,2-dibromo-3-chloropropane, (H) methyl tert-butyl ether, (I) 1,1-dichloroethene, (J) CFC-11, (K) CFC-12, (L) atrazine, (M) simazine, (N) bromacil, (O) diuron, and (P) perchlorate, Upper Santa Ana Watershed study unit, GAMA Priority Basin Project.

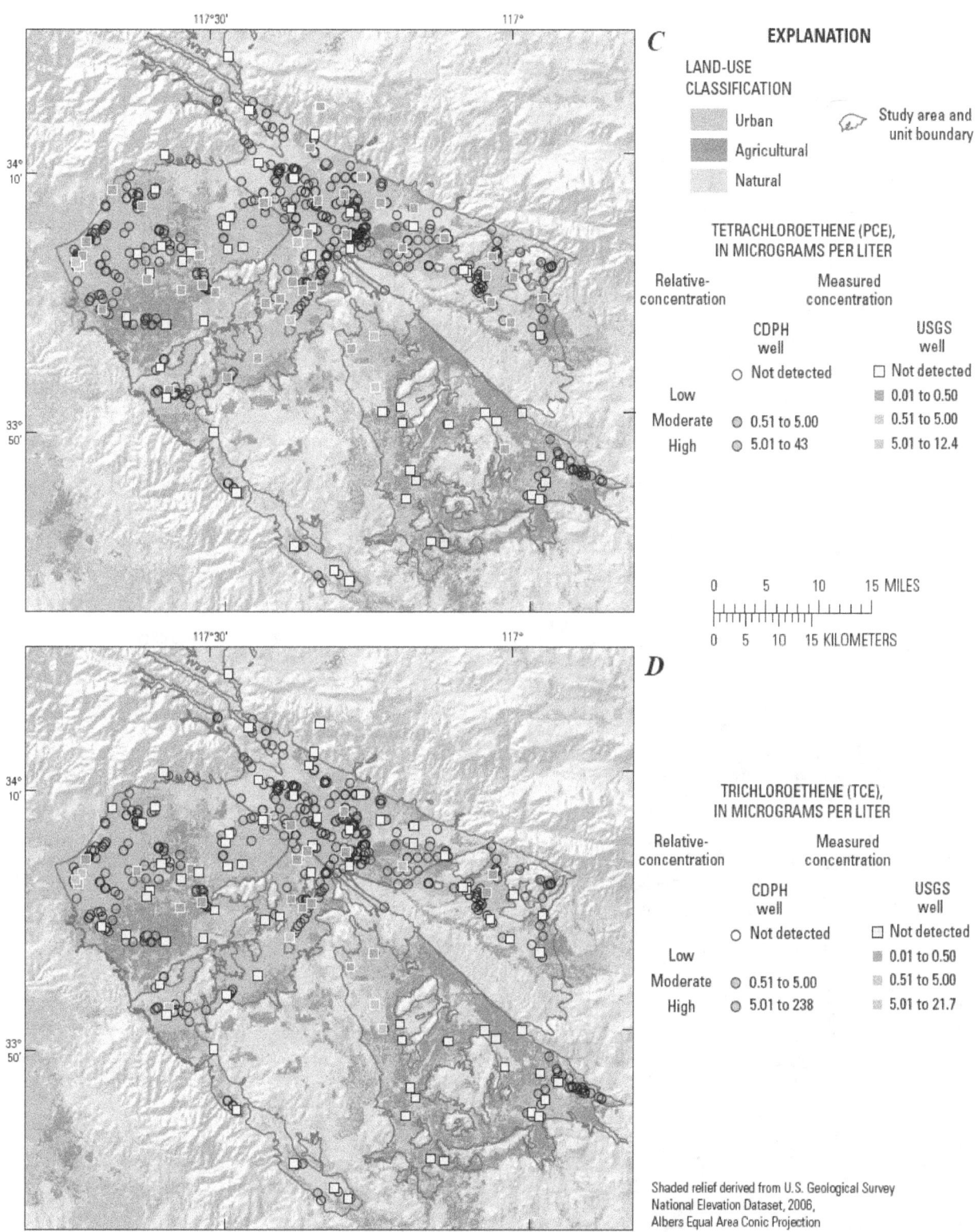

C

EXPLANATION

LAND-USE
CLASSIFICATION

Urban

Agricultural

Natural

Study area and
unit boundary

TETRACHLOROETHENE (PCE),
IN MICROGRAMS PER LITER

Relative-
concentration

Measured
concentration

CDPH
well

USGS
well

Not detected (○)   Not detected (□)

Low   0.01 to 0.50

Moderate   0.51 to 5.00   0.51 to 5.00

High   5.01 to 43   5.01 to 12.4

0   5   10   15 MILES

0   5   10   15 KILOMETERS

D

TRICHLOROETHENE (TCE),
IN MICROGRAMS PER LITER

Relative-
concentration

Measured
concentration

CDPH
well

USGS
well

Not detected (○)   Not detected (□)

Low   0.01 to 0.50

Moderate   0.51 to 5.00   0.51 to 5.00

High   5.01 to 238   5.01 to 21.7

Shaded relief derived from U.S. Geological Survey
National Elevation Dataset, 2006,
Albers Equal Area Conic Projection

Figure 12.—Continued.

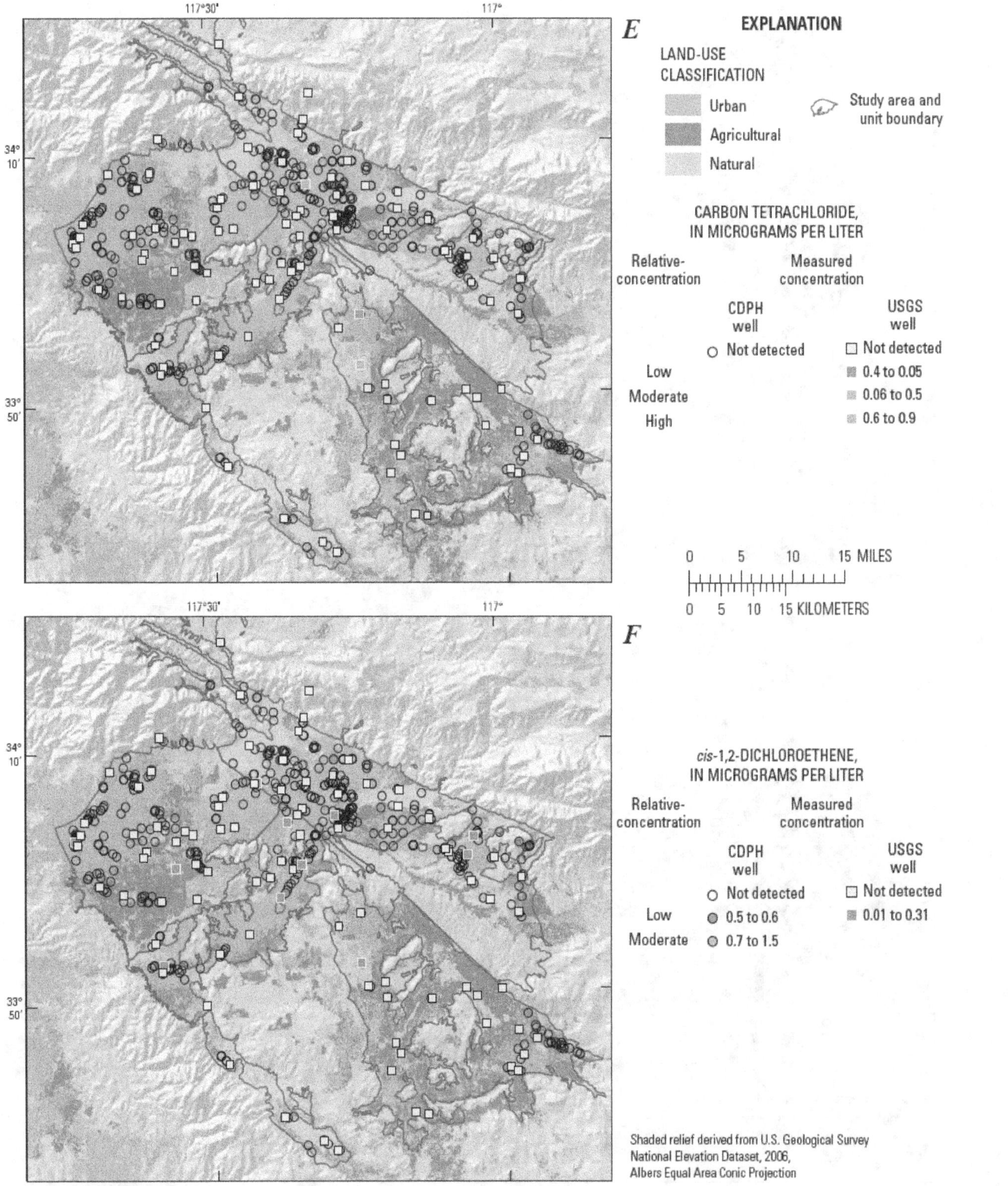

117°30'
117°
34° 10'
33° 50'
E
EXPLANATION
LAND-USE CLASSIFICATION
Urban
Agricultural
Natural
Study area and unit boundary
CARBON TETRACHLORIDE, IN MICROGRAMS PER LITER
Relative-concentration
Measured concentration
CDPH well
USGS well
Not detected
Not detected
Low
0.4 to 0.05
Moderate
0.06 to 0.5
High
0.6 to 0.9
0 5 10 15 MILES
0 5 10 15 KILOMETERS
F
cis-1,2-DICHLOROETHENE, IN MICROGRAMS PER LITER
Relative-concentration
Measured concentration
CDPH well
USGS well
Not detected
Not detected
Low
0.5 to 0.6
0.01 to 0.31
Moderate
0.7 to 1.5
Shaded relief derived from U.S. Geological Survey National Elevation Dataset, 2006, Albers Equal Area Conic Projection

Figure 12.—Continued.

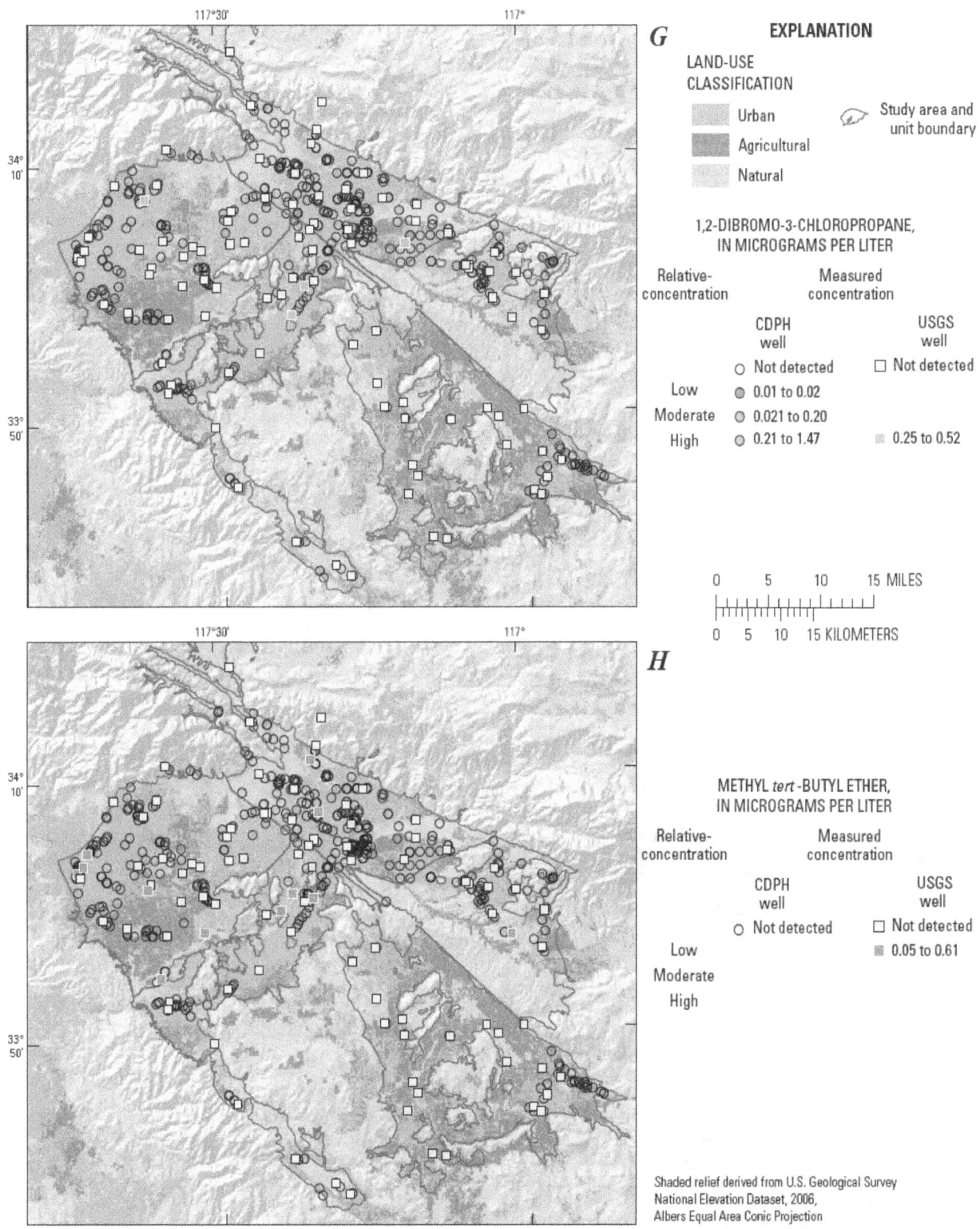

EXPLANATION

G

LAND-USE
CLASSIFICATION

Urban

Agricultural

Natural

Study area and
unit boundary

1,2-DIBROMO-3-CHLOROPROPANE,
IN MICROGRAMS PER LITER

Relative-
concentration

Measured
concentration

CDPH
well

USGS
well

Not detected

Not detected

Low   0.01 to 0.02

Moderate   0.021 to 0.20

High   0.21 to 1.47   0.25 to 0.52

0   5   10   15 MILES

0   5   10   15 KILOMETERS

H

METHYL *tert*-BUTYL ETHER,
IN MICROGRAMS PER LITER

Relative-
concentration

Measured
concentration

CDPH
well

USGS
well

Not detected

Not detected

Low   0.05 to 0.61

Moderate

High

Shaded relief derived from U.S. Geological Survey
National Elevation Dataset, 2006,
Albers Equal Area Conic Projection

Figure 12.—Continued.

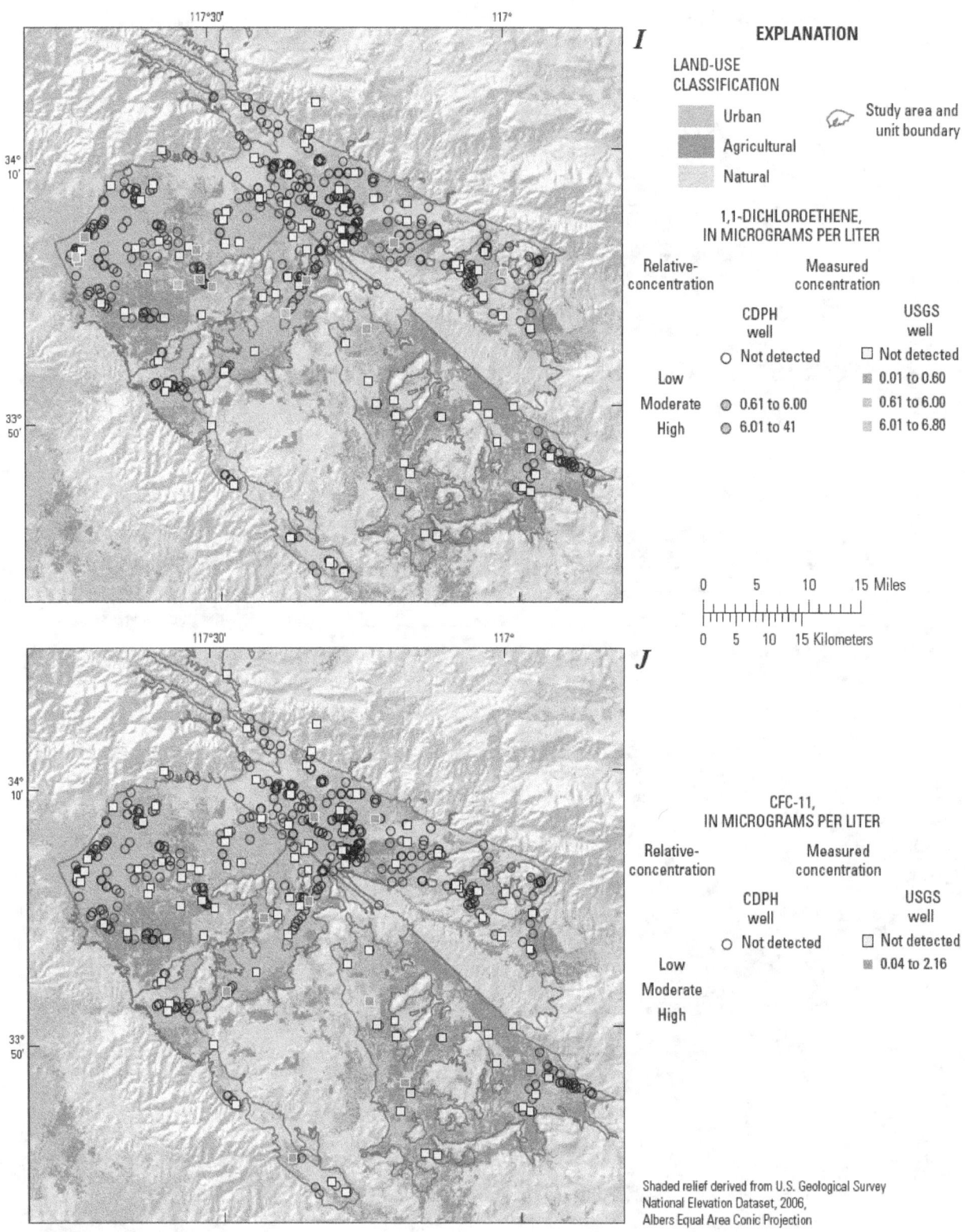

EXPLANATION

LAND-USE CLASSIFICATION

Urban

Agricultural

Natural

Study area and unit boundary

1,1-DICHLOROETHENE, IN MICROGRAMS PER LITER

Relative-concentration	Measured concentration	
	CDPH well	USGS well
	○ Not detected	□ Not detected
Low		▣ 0.01 to 0.60
Moderate	◉ 0.61 to 6.00	▣ 0.61 to 6.00
High	◉ 6.01 to 41	▣ 6.01 to 6.80

0   5   10   15 Miles

0   5   10   15 Kilometers

CFC-11, IN MICROGRAMS PER LITER

Relative-concentration	Measured concentration	
	CDPH well	USGS well
	○ Not detected	□ Not detected
Low		▣ 0.04 to 2.16
Moderate		
High		

Shaded relief derived from U.S. Geological Survey National Elevation Dataset, 2006, Albers Equal Area Conic Projection

Figure 12.—Continued.

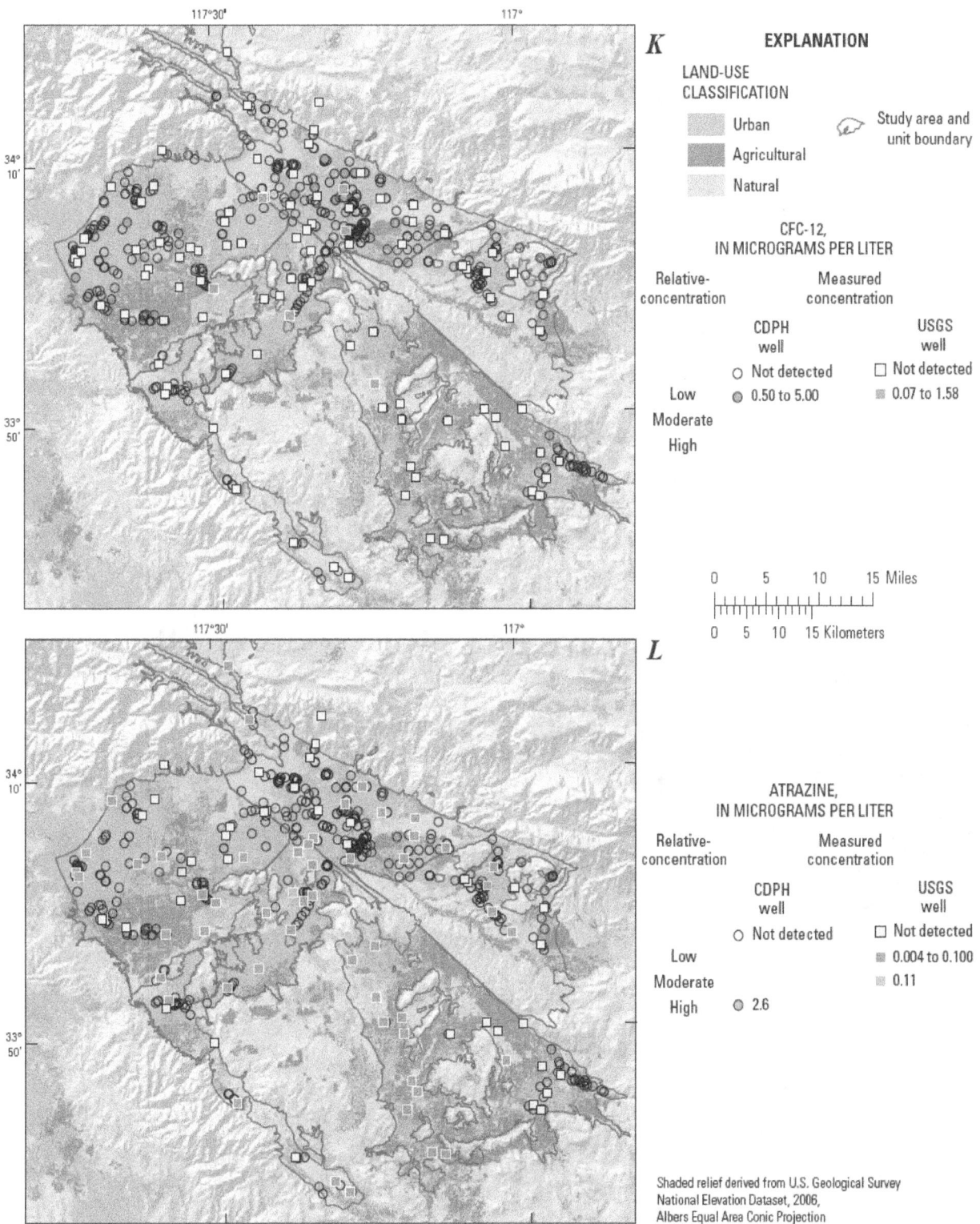

Figure 12.—Continued.

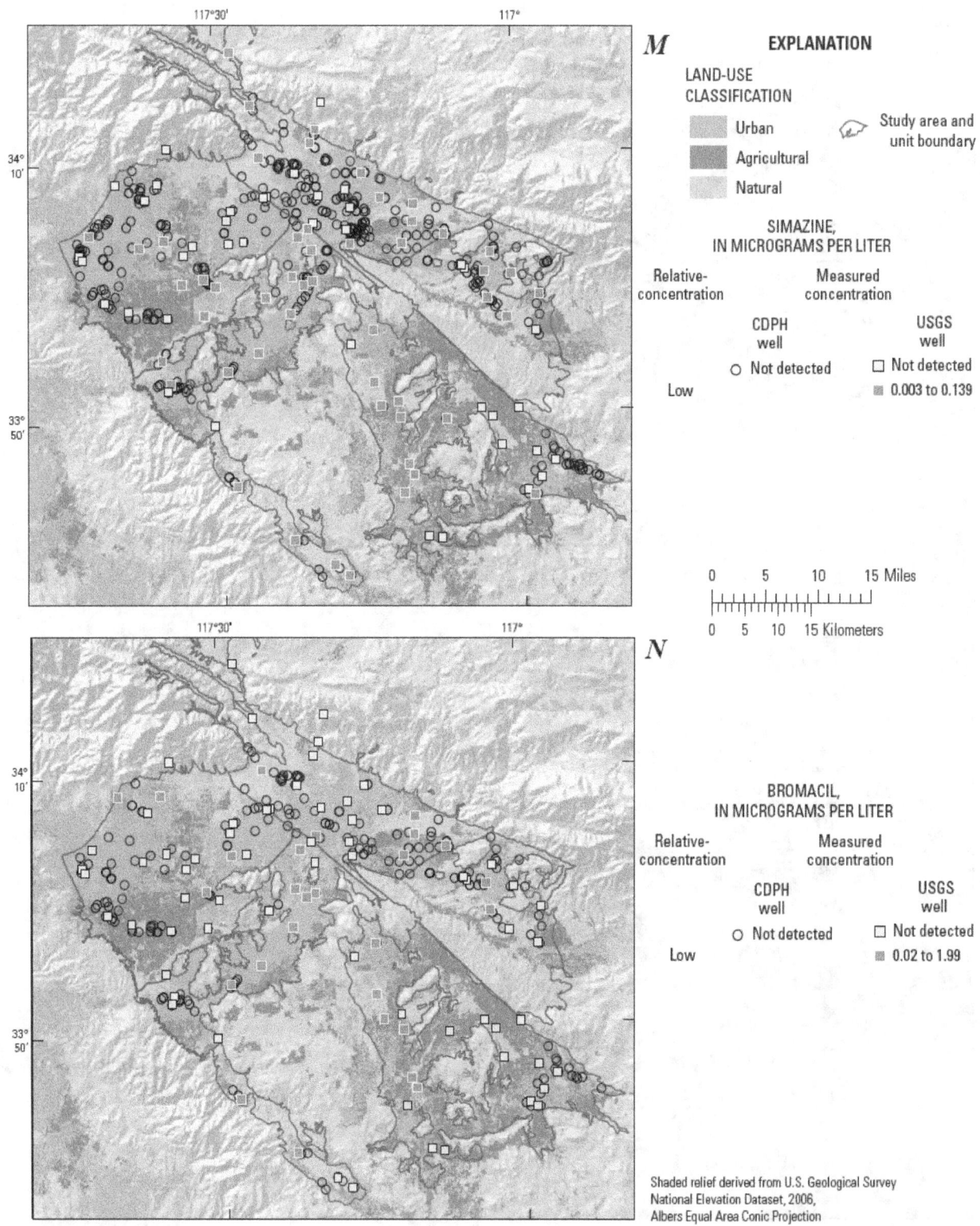

Figure 12.—Continued.

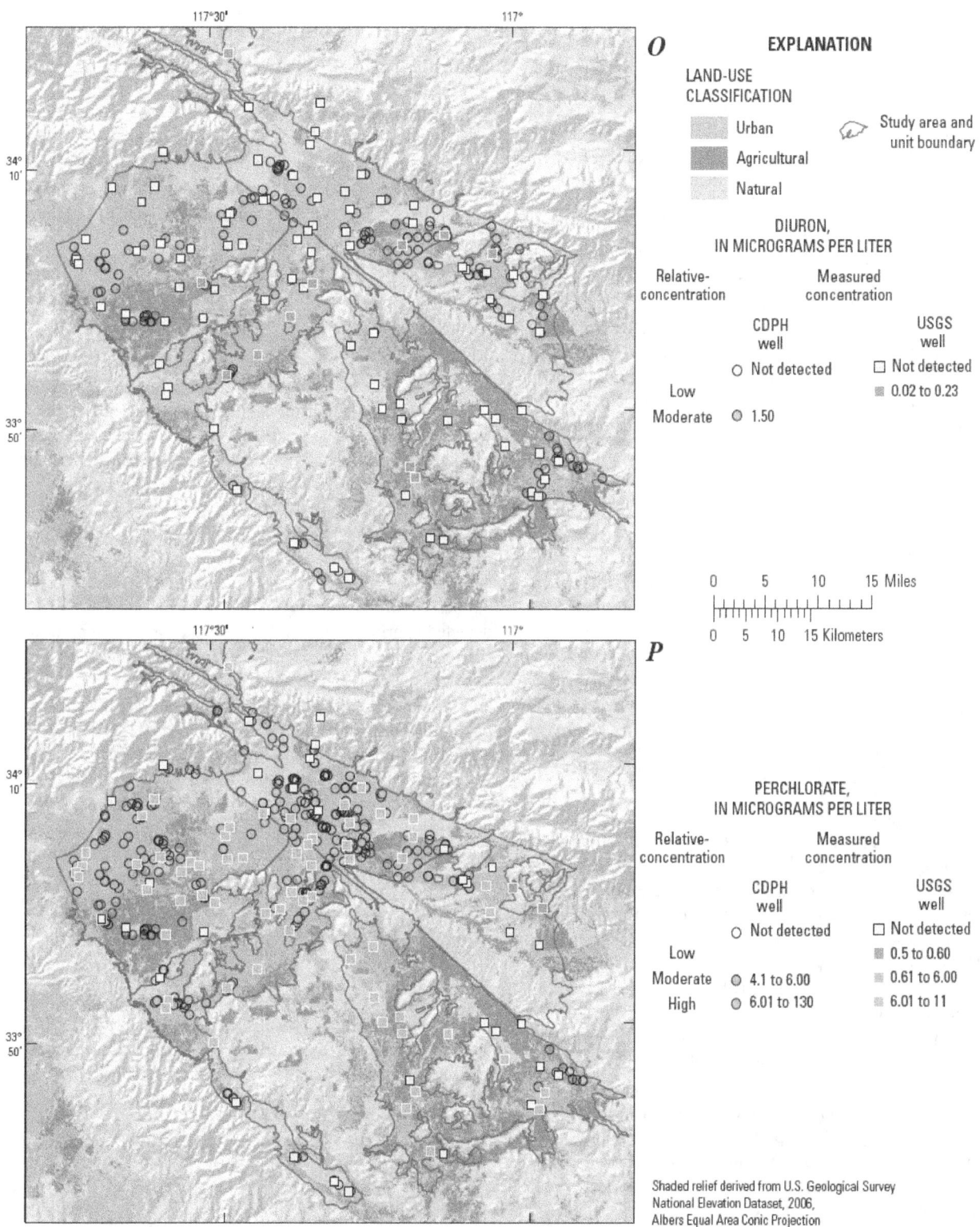

EXPLANATION

LAND-USE
CLASSIFICATION

Urban

Agricultural

Natural

Study area and
unit boundary

DIURON,
IN MICROGRAMS PER LITER

Relative-
concentration

Measured
concentration

CDPH
well

USGS
well

○ Not detected

□ Not detected

Low

▦ 0.02 to 0.23

Moderate   ◉ 1.50

0   5   10   15 Miles

0   5   10   15 Kilometers

PERCHLORATE,
IN MICROGRAMS PER LITER

Relative-
concentration

Measured
concentration

CDPH
well

USGS
well

○ Not detected

□ Not detected

Low

▦ 0.5 to 0.60

Moderate   ◉ 4.1 to 6.00

▦ 0.61 to 6.00

High   ◉ 6.01 to 130

▦ 6.01 to 11

Shaded relief derived from U.S. Geological Survey
National Elevation Dataset, 2006,
Albers Equal Area Conic Projection

Figure 12.—Continued.

PCE and TCE

PCE and TCE were detected in all USAW study areas except Elsinore, and PCE was detected at a high relative-concentration in one grid well in each of the following study areas: Cucamonga/Chino, Riverside-Arlington/Temescal, and San Jacinto (figs. 11, 12C). The grid wells with high relative-concentrations of PCE in the Cucamonga/Chino and Riverside-Arlington/Temescal study areas also had high relative-concentrations of TCE. The San Jacinto grid well with a high relative-concentration of PCE had a moderate relative-concentration of TCE (fig. 12D). High and moderate relative-concentrations of PCE and TCE in CDPH wells generally were detected in the same areas where they were detected in USAW grid wells, with a few exceptions. For example, high and moderate relative-concentrations of TCE were detected in CDPH wells in the southern portion of the Cucamonga/Chino study area. TCE was not detected in USAW grid wells in this area.

Fumigants

Fumigants are used as agricultural or household pesticides. Of the nine fumigants (table 3A, Kent and Belitz, 2009) sampled for in this study, two—1,2-dichloropropane and 1,2 dibromo-3-chloropropane (DBCP)—were detected in USAW grid wells.

DBCP

DBCP was detected in four USGS grid wells, and all four detections were at concentrations above the USEPA MCL. The four detections at high relative-concentrations resulted in a grid-based high aquifer-scale proportion of 4.4 percent for this compound (table 5). Two of the USGS grid-well detections occurred in wells located in the Riverside-Arlington/Temescal study area. The other two USGS grid-well detections occurred in one well each of the Cucamonga/Chino and Bunker Hill/Cajon/Rialto-Colton study areas (figs. 11, 12G). Detections of DBCP in CDPH wells were in these same general locations. The analytical method used by CDPH had a lower detection limit than that used by the USGS, allowing detections of DBCP at low and moderate relative-concentrations for CDPH wells in the study unit. The spatially weighted moderate aquifer proportion was 4.1 percent; the spatially weighted high aquifer proportion of 3.9 percent was similar to the grid-based aquifer-scale proportion of 4.4 percent and within the 90 percent confidence interval of the grid-based estimate (table 5). Fumigants that have been historically, though not currently, high in the USAW study unit include the previously mentioned 1,2-dichloropropane and 1,2-dibromoethane (table 4).

Other VOCs

In this report, the 43 compounds under the category "other VOCs" (any VOC other than a THM, solvent, or fumigant) include fuel components, fire retardants, refrigerants, and compounds used in the synthesis of organic chemicals. Four compounds in the category "other VOCs" met the selection criteria for further discussion: the fuel component MTBE, the organic synthesis compound 1,1-DCE, and the refrigerants CFC-11 and CFC-12.

MTBE is a compound that was used to oxygenate gasoline until its use was discontinued in California after December 2003 (Rausser and others, 2004). MTBE was detected at low relative-concentrations in 11 percent of USAW grid wells (table 5) and was detected in all of the USAW study areas except for San Jacinto and Elsinore (figs. 11, 12H). While all current and historical detections of MTBE in the USAW study unit have been at low relative-concentrations, a similar compound—*tert*-butyl alcohol, classified as a gasoline oxygenate degradate—was detected in a well at a high relative-concentration prior to November 2003 (table 4).

The organic synthesis compound 1,1-DCE was detected in 12 percent of the primary aquifers. Relative-concentrations of 1,1-DCE were high in 1.1 percent and moderate in 3.3 percent of the primary aquifers (table 5). 1,1-DCE was detected in grid wells of all USAW study areas except for the Elsinore study area (fig. 12I). The single detection at a high relative-concentration occurred in the Cucamonga/Chino study area, and moderate relative-concentrations were detected in grid wells of the Cucamonga/Chino, Riverside-Arlington/Temescal, and Yucaipa/San Timoteo study areas (figs. 11, 12I). A few additional detections of 1,1-DCE at high and moderate relative-concentrations occurred in CDPH wells that were not grid wells (fig.12I). Two additional organic synthesis compounds, chloromethane and vinyl chloride, had high relative-concentrations in at least one well in the CDPH data before November 2003 (table 4). Neither of these compounds was detected in USAW grid wells.

Two refrigerant compounds, both chlorofluorocarbons, were detected at low relative-concentrations in at least 10 percent of the grid wells. CFC-11 was detected in 11 percent of USAW grid wells (table 5) and was detected in all of the study areas except for the Yucaipa/San Timoteo and the Cucamonga Chino study areas (figs. 11, 12J). Dichlorodifluoromethane (CFC-12) was detected in 10 percent (9 of 90) of the grid wells (table 5) and was detected in all of the study areas except for the Yucaipa/San Timoteo and Elsinore study areas (figs. 11, 12K).

Herbicides

Four herbicides—atrazine, simazine, bromacil, and diuron—were detected in 10 percent or more of the grid wells which, for this report, defines "frequently detected." At least one herbicide was detected in 68 percent of the 90 grid wells. The largest relative-concentration for a herbicide was 0.11 (moderate) for atrazine. Atrazine was also the herbicide most frequently detected in USAW grid wells (62 percent), followed closely by simazine (54 percent) (table 5; fig. 11); these two herbicides were frequently detected in all of the USAW study areas (fig. 12L, 12M). Detections of bromacil and diuron were at low relative-concentrations (fig. 11). Bromacil was detected in all of the USAW study areas except for the Yucaipa/San Timoteo study area (figs. 11, 12N). Diuron was detected in all of the USAW study areas except for the Elsinore study area (figs. 11, 12O). Hamlin and others (2005) reported concentrations and detection frequencies for herbicides similar to those in this study in samples collected from 1999 to 2001 as part of a USGS National Water-Quality Assessment Program study carried out in the Santa Ana Basin. Atrazine was detected at a high relative concentration in one CDPH well in the Yucaipa\San Timoteo study area (fig. 12L).

Insecticides and Fungicides

None of the 35 insecticides or 21 insecticide degradates analyzed for in this study were detected (Kent and Belitz, 2009). Relative concentrations cannot be calculated for metalaxyl because it does not have a benchmark. A fungicide, metalaxyl, was detected at low concentrations in two grid wells (Kent and Belitz, 2009). There were no detections of the other seven fungicides analyzed for in this study, nor of the single fungicide degradate, 3,4-dichloroaniline. Historically, the only pesticide with high relative-concentrations in the CDPH data before November 2003 was the insecticide heptachlor (table 4).

Special-Interest Constituents

Constituents of special interest analyzed for in the USAW study unit were 1,4-dioxane, N-nitrosodimethylamine (NDMA), 1,2,3-trichloropropane (1,2,3-TCP), and perchlorate. These constituents were selected because they recently have been detected in, or are considered to have the potential to reach, drinking-water supplies (California Department of Public Health, 2008b,c,d). NDMA is a highly toxic byproduct of the chlorination of wastewater (Bradley and others, 2005), and prior to April 1976, it was produced as a component of rocket fuel (Agency for Toxic Substances and Disease Registry, 1989). 1,4-Dioxane is a compound used as a stabilizer for chlorinated solvents (Tilman, 2009). NDMA and 1,4-dioxane were not detected in any samples (Kent and Belitz, 2009).

1,2,3-TCP is sometimes classified as a solvent (Bender and others, 1999), or as a fumigant (Oki and Giambelluca, 1987; Zebarth and others, 1998; Zogorski and others, 2006; Landon and others, 2010), reflecting spatial variations in its predominant use. The classification of 1,2,3-TCP is of little importance here because its occurrence did not meet the selection criteria for additional evaluation in this report. 1,2,3-TCP was detected at low relative-concentrations in 9.4 percent of the grid wells.

Perchlorate

Most perchlorate found in groundwater has been attributed to its use as an oxidizer in solid propellants for rockets, fireworks, and other explosives (Orris and others, 2003). Perchlorate also has natural sources, such as Chilean caliche, is used as a nitrate fertilizer (Urbansky and others, 2001), and it can be present at low concentrations in groundwater under natural conditions (Fram and Belitz, 2011). Perchlorate was detected in 67 percent of the grid wells, had a high aquifer proportion of 11.1 percent, and had a moderate aquifer proportion of 53.3 percent (table 5). Perchlorate was detected at high and moderate relative-concentrations in all USAW study areas except for the Elsinore study area; however, high relative-concentrations were most prevalent in the Riverside-Arlington/Temescal study area (figs. 11, 12P). The Yucaipa/San Timoteo study area had moderate, but no high relative-concentrations in grid wells.

Summary

Groundwater quality in the approximately 1,000-square-mile (2,590-square-kilometer) Upper Santa Ana Watershed (USAW) study unit was investigated as part of the Priority Basin Project of the Groundwater Ambient Monitoring and Assessment (GAMA) Program. Samples were collected during November 2006 through March 2007 from 99 wells in 6 study areas: Bunker Hill/Cajon/Rialto-Colton, Cucamonga/Chino, Riverside-Arlington/Temescal, Yucaipa/San Timoteo, San Jacinto, and Elsinore. The GAMA Priority Basin Project is designed to provide a statistically robust characterization of untreated-groundwater quality in the primary aquifers at the basin-scale. Ninety wells were randomly selected within spatially distributed grid cells across the USAW study unit to assess the quality of the groundwater. An additional nine wells were sampled for the purposes of better understanding the relation of water quality to explanatory factors. Samples from USGS-grid wells were analyzed for up to 318 constituents. CDPH inorganic data from the prior 3-year period (November 30, 2003, to December 1, 2006) were used to complement USGS data and provide additional information about groundwater quality.

Relative-concentrations (sample concentration divided by the health- or aesthetic-based benchmark concentration) were used for evaluating groundwater quality for those constituents that have Federal and (or) California regulatory or non-regulatory benchmarks for drinking-water quality.

Aquifer-scale proportion was used as the primary metric for evaluating regional-scale groundwater quality. High aquifer-scale proportion is defined as the percentage of the primary aquifers with relative-concentration greater than 1.0 for a particular constituent or class of constituents; proportion is based on an areal rather than a volumetric basis. Moderate and low aquifer-scale proportions were defined as the percentage of the primary aquifers with moderate and low relative-concentrations, respectively. Two statistical approaches, grid-based and spatially weighted, were used to evaluate aquifer-scale proportions for individual constituents and classes of constituents. Grid-based and spatially weighted estimates were comparable in the USAW study unit (within 90 percent confidence intervals). However, the spatially weighted approach was superior to the grid-based proportion when the relative concentration of a constituent was high in a small fraction of the aquifer.

Inorganic constituents with human-health benchmarks were detected at high relative-concentrations in 32.9 percent of the primary aquifers, moderate in 29.3 percent, and low in 37.8 percent. The high aquifer-scale proportion of inorganic constituents primarily reflected high aquifer-scale proportions of nitrate plus nitrite (25.3 percent), gross alpha activity (5.6 percent), arsenic (5.1 percent), molybdenum (4.1 percent), and boron (3.0 percent).

Inorganic constituents with aesthetic benchmarks and secondary maximum contaminant levels (SMCLs) were detected at high relative-concentrations in 8.8 percent of the primary aquifers, and at moderate relative-concentrations in 28.8 percent. Total dissolved solids (TDS) was the inorganic constituent with an SMCL that most frequently had high relative-concentrations (4.9 percent).

Of the 84 volatile organic compounds (VOCs) analyzed, 21 were detected (this excludes 1,2,3-trichloropropane, considered separately as a constituent of special interest). All 21 of the VOCs detected had human-health benchmarks, and 5 were detected at high relative-concentration in at least one sample. These were the solvents tetrachloroethene, trichloroethene, and carbon tetrachloride, the fumigant 1,2-dibromo-3-chloropropane, and the organic synthesis compound 1,1-dichloroethene. The two isomers of dichloroethane (1,1- and 1,2-) each had moderate relative-concentrations in 1.1 percent of the primary aquifers. The remaining VOCs that were detected were detected at only low relative-concentrations. Six of these were detected in 10 percent or more of the grid wells. These were the solvent cis-1,2-dichloroethene, the trihalomethanes chloroform and bromodichloromethane, the refrigerants CFC-11 and CFC-12, and the gasoline oxygenate methyl tert-butyl ether.

Of the 135 pesticide compounds analyzed, 13 were detected. Seven of these 13 had human-health benchmarks. Pesticides did not have high relative-concentrations in any proportion of the primary aquifer system. The pesticide, atrazine, had a moderate relative-concentration in 1.1 percent of the grid wells, and was detected in 62 percent of the wells. Three pesticides, simazine, bromacil, and diuron, were detected at low relative-concentrations in 54 percent, 19 percent, and 12 percent of the grid wells, respectively.

Four constituents of special interest were analyzed for in the USAW study unit: 1,4-dioxane, N-nitrosodimethylamine (NDMA), 1,2,3-trichloropropane (1,2,3-TCP), and perchlorate. NDMA and 1,4-dioxane were not detected in any samples, and 1,2,3-TCP was detected only at low relative-concentrations in 9.4 percent of the grid wells. In contrast, perchlorate was detected in 67 percent of the primary aquifers. Perchlorate was detected at high relative-concentrations in 11.1 percent of the primary aquifers, and at moderate relative-concentrations in 53.3 percent.

Acknowledgments

The authors thank the following cooperators for their support: the California State Water Resources Control Board, Lawrence Livermore National Laboratory, the California Department of Public Health, and the California Department of Water Resources. We especially thank the cooperating well owners and water purveyors for their cooperation in allowing the USGS to collect samples from their wells. Funding for this work was provided by State of California bonds authorized by Proposition 50 and administered by the State Water Board.

References

Agency for Toxic Substances and Disease Registry, 1989, Toxicological profile for N-Nitrosodimethylamine: Agency for Toxic Substances and Disease Registry, U.S. Public Health Service, in collaboration with the U.S. Environmental Protection Agency, 120 p., accessed September 16, 2011, at http://www.atsdr.cdc.gov/toxprofiles/tp141.pdf.

Arndt, M.F., 2010, Evaluation of gross alpha and uranium measurements for MCL compliance: Denver, CO, Water Research Foundation, 299 p.

Ayotte, J.D., Szabo, Z., Focazio, M.J., and Eberts, S.M., 2011, Effects of human-induced alteration of groundwater flow on concentrations of naturally-occurring trace elements at water-supply wells: Applied Geochemistry, v. 26, no. 5, p. 747–762, doi:101016/j.apgeochem.2011.01.033.

Belitz, K., Dubrovsky, N.M., Burow, K.R., Jurgens, B., and Johnson, T., 2003, Framework for a groundwater quality monitoring and assessment program for California: U.S. Geological Survey Water-Resources Investigations Report 03-4166, 28 p.

Belitz, K., Hamlin, S., Burton, C., Kent, R., Fay, R.G., and Johnson, T., 2004, Water quality in the Santa Ana Basin, California, 1999–2001: U.S. Geological Survey Circular 1238, 37 p.

Belitz, K., Jurgens, B., Landon, M.K., Fram, M.S., and Johnson, T., 2010, Estimation of aquifer-scale proportion using equal-area grids—Assessment of regional-scale groundwater quality: Water Resources Research, v. 46, W11550, 14 p., doi:10.1029/2010WR009321. (Also available at http://www.agu.org/pubs/crossref/2010/2010WR009321.shtml.)

Bender, D.A., Zogorski, J.S., Halde, M.J., and Rowe, B.L., 1999, Selection procedure and salient information for volatile organic compounds emphasized in the National Water-Quality Assessment Program: U.S. Geological Survey Open-File Report 99–182, 32 p.

Bradley, P.M., Carr, S.A., Baird, R.B., and Chapelle, F.H., 2005, Bieodegradation of N-nitrosodimethylamine in soil from a water reclamation facility: Bioremediation Journal, v. 9, no. 2, p. 115–120.

Brown, L.D., Cai, T.T., and Dasgupta, A., 2001, Interval estimation for a binomial proportion: Statistical Science, v. 16, no. 2, p. 101–117.

California Department of Finance, 2000, Demographic Research Unit, 1970–1980–1990–2000 Comparability File, digital dataset of US CENSUS data normalized to the 1990 tract boundaries for the State of California.

California Department of Public Health, 2008a, Maximum contaminant levels and regulatory dates for drinking water, U.S. EPA vs. California, accessed September 23, 2011, at http://www.cdph.ca.gov/certlic/drinkingwater/Documents/DWdocuments/EPAandCDPH-11-28-2008.pdf.

California Department of Public Health, 2008b, Perchlorate in drinking water, accessed June 2, 2010, at http://www.cdph.ca.gov/certlic/drinkingwater/Pages/Perchlorate.aspx.

California Department of Public Health, 2008c, California drinking water—NDMA- related activities, accessed June 2, 2010, at http://www.cdph.ca.gov/certlic/drinkingwater/Pages/NDMA.aspx.

California Department of Public Health, 2008d, 1,2,3-Trichloropropane, accessed June 2, 2010, at http://www.cdph.ca.gov/certlic/drinkingwater/Pages/123TCP.aspx.

California Department of Public Health, 2012, California Code of Regulation, Title 22, Division 4 Environmental Health, Chapter 15 Domestic water quality and monitoring regulations, accessed February 2012 at http://ccr.oal.ca.gov.

California Department of Water Resources, 2003, California's groundwater: California Department of Water Resources Bulletin 118, 246 p., accessed September 9, 2011, at http://www.water.ca.gov/groundwater/bulletin118/update2003.cfm.

California Department of Water Resources, 2004a, Upper Santa Ana Valley groundwater basin, Bunker Hill subbasin: California's Groundwater Bulletin 118, accessed September 9, 2011, at http://www.water.ca.gov/pubs/groundwater/bulletin_118/basindescriptions/8-2.06.pdf.

California Department of Water Resources, 2004b, Upper Santa Ana Valley groundwater basin, Cajon subbasin: California's Groundwater Bulletin 118, accessed September 9, 2011, at http://www.water.ca.gov/pubs/groundwater/bulletin_118/basindescriptions/8-2.05.pdf.

California Department of Water Resources, 2004c, Upper Santa Ana Valley groundwater basin, Rialto-Colton subbasin: California's Groundwater Bulletin 118, accessed September 9, 2011, at http://www.water.ca.gov/pubs/groundwater/bulletin_118/basindescriptions/8-2.04.pdf.

California Department of Water Resources, 2004d, Upper Santa Ana Valley groundwater basin, Cucamonga subbasin: California's Groundwater Bulletin 118, accessed September 9, 2011, at http://www.water.ca.gov/pubs/groundwater/bulletin_118/basindescriptions/8-2.02.pdf.

California Department of Water Resources, 2004e, Upper Santa Ana Valley groundwater basin, Yucaipa Subbasin: California's Groundwater Bulletin 118, accessed September 9, 2011, at http://www.water.ca.gov/pubs/groundwater/bulletin_118/basindescriptions/8-2.07.pdf.

California Department of Water Resources, 2004f, Upper Santa Ana Valley groundwater basin, San Timoteo Subbasin: California's Groundwater Bulletin 118, accessed September 9, 2011, at http://www.water.ca.gov/pubs/groundwater/bulletin_118/basindescriptions/8-2.08.pdf.

California Department of Water Resources, 2004g, Upper Santa Ana Valley groundwater basin, Riverside-Arlington Subbasin: California's Groundwater Bulletin 118, accessed September 9, 2011, at http://www.water.ca.gov/pubs/groundwater/bulletin_118/basindescriptions/8-2.03.pdf.

California Department of Water Resources, 2006a, Upper Santa Ana Valley groundwater basin, Chino Subbasin: California's Groundwater Bulletin 118, accessed September 9, 2011, at http://www.water.ca.gov/pubs/groundwater/bulletin_118/basindescriptions/8-2.01.pdf.

California Department of Water Resources, 2006b, Upper Santa Ana Valley groundwater basin: Temescal Subbasin: California's Groundwater Bulletin 118, accessed September 9, 2011, at http://www.water.ca.gov/pubs/groundwater/bulletin_118/basindescriptions/8-2.09.pdf.

California Department of Water Resources, 2006c, Elsinore Groundwater Basin, California's Groundwater Bulletin 118, accessed September 9, 2011, at http://www.water.ca.gov/pubs/groundwater/bulletin_118/basindescriptions/8-4.pdf.

California Department of Water Resources, 2006d, San Jacinto Groundwater Basin, California's Groundwater Bulletin 118, accessed September 9, 2011, at http://www.water.ca.gov/pubs/groundwater/bulletin_118/basindescriptions/8-5.pdf.

California Department of Water Resources, Southern District, 1981, Investigation of ground water supply for stabilization of level of Lake Elsinore, Riverside County: District Report to Department of Parks and Recreation under Interagency agreement 162083, 45 p.

California State Water Resources Control Board, 2003, A comprehensive groundwater quality monitoring program for California: Assembly Bill 599 Report to the Governor and Legislature, March 2003, 100 p., accessed January 11, 2012, at http://www.waterboards.ca.gov/gama/docs/final_ab_599_rpt_to_legis_7_31_03.pdf.

California State Water Resources Control Board, 2012, GAMA – Groundwater Ambient Monitoring & Assessment Program: California Department of Environmental Protection website, accessed January 25, 2012, at http://www.waterboards.ca.gov/gama/.

Chapelle, F.H., 2001, Groundwater microbiology and geochemistry (2[d] ed.): New York, John Wiley and Sons, Inc., 477 p.

Chapelle, F.H., McMahon, P.B., Dubrovsky, N.M., Fuji, R.F., Oaksford, E.T., and Vroblesky, D.A., 1995, Deducing the distribution of terminal electron-accepting processes in hydrologically diverse groundwater systems: Water Resources Research, v. 31, no. 2, p. 359–371.

Childress, C.J.O., Foreman, W.T., Connor, B.F., and Maloney, T.J., 1999, New reporting procedures based on long-term method detection levels and some considerations for interpretations of water-quality data provided by the U.S. Geological Survey National Water Quality Laboratory: U.S. Geological Survey Open-File Report 99–193, 19 p.

Clark, I.D., and Fritz, P., 1997, Environmental isotopes in hydrogeology: New York, Lewis Publishers, 328 p.

Cook, P.G., and Böhlke, J.K., 2000, Determining timescales for groundwater flow and solute transport, *in* Cook, P.G., and Herczeg, A., eds., Environmental tracers in subsurface hydrology: Boston, Kluwer Academic Publishers, p. 1–30.

Craig, H., and Lal, D., 1961, The production rate of natural tritium: Tellus, v. 13, p. 85–105.

Danskin, W.R., McPherson, K.R., and Woolfenden, L.R., 2006, Hydrology, description of computer models, and evaluation of selected water-management alternatives in the San Bernardino area, California: U.S. Geological Survey Open-File Report 2005–1278, 178 p., 2 pl.

Dawson, B.J.M., Belitz, K., Land, M.T., and Danskin, W.R., 2003, Stable isotopes and volatile organic compounds along seven groundwater flow paths in divergent and convergent flow systems, southern California, 2000: U.S. Geological Survey Water-Resources Investigations Report 03–4059, 79 p. Available at http://water.usgs.gov/pubs/wri/wrir034059/.

Dutcher, L.C., and Garrett, A.A., 1963, Geologic and hydrologic features of the San Bernardino area, California—with special reference to underflow across the San Jacinto Fault: U.S. Geological Survey Water-Supply Paper 1419, 114 p.

Eastern Municipal Water District, 2002, Regional groundwater model for the San Jacinto watershed: Techlink Environmental, Inc. [variously paged].

Faye, R.E., 1973, Groundwater hydrology of northern Napa Valley, California: U.S. Geological Survey Water-Resources Investigations 13–73, 64 p.

Fontes, J.C., and Garnier, J.M., 1979, Determination of the initial ^{14}C activity of the total dissolved carbon—A review of the existing models and a new approach: Water Resources Research, v. 15, no. 2, p. 399–413.

Fram, M.S., and Belitz, Kenneth, 2011, Occurrence and concentrations of pharmaceutical compounds in groundwater used for public drinking-water supply in California: Science of the Total Environment, v. 409, no. 18, p. 3409–3417, doi:10.1016/j.scitotenv.2011.05.053. Available at http://ca.water.usgs.gov/news/2011/ReleaseJuly05_2011-Fram_Belitz.pdf.

Fram, M.S., and Belitz, Kenneth, 2011, Probability of detecting perchlorate under natural conditions in deep groundwater in California and the Southwestern United States: Environmental Science and Technology, January 2011. Available at http://pubs.acs.org/doi/pdfplus/10.1021/es103103p.

Gilliom, R.J., Barbash, J.E., Crawford, C.G., Hamilton, P.A., Martin, J.D., Nakagaki, N., Nowell, L.H., Scott, J.C., Stackelberg, P.E., Thelin, G.P., and Wolock, D.M., 2006, The quality of our nation's waters: Pesticides in the nation's streams and ground water, 1992–2001: U.S. Geological Survey Circular 1291, 172 p.

Hamlin, S.N., Belitz, Kenneth, and Johnson, T., 2005, Occurrence and distribution of volatile organic compounds and pesticides in ground water in relation to hydrogeologic characteristics and land use in the Santa Ana Basin, southern California: U.S. Geological Survey Scientific Investigations Report 2006–5032, 40 p.

Hamlin, S.N., Belitz, Kenneth, Kraja, S., and Dawson, B.J., 2002, Groundwater quality in the Santa Ana Watershed, California—Overview and data summary: U.S. Geological Survey Water-Resources Investigations Report 02–4243, 137 p.

Hem, J.D., 1992, Study and interpretation of the chemical characteristics of natural water, third edition: U.S. Geological Survey Water Supply Paper 2254, 264 p.

Isaaks, E.H., and Srivastava, R.M., 1989, Applied geostatistics: New York, Oxford University Press, 511 p.

Ivahnenko, Tammy, and Barbash, J.E., 2004, Chloroform in the hydrologic system—Sources, transport, fate, occurrence, and effects on human health and aquatic organisms: U.S. Geological Survey Scientific Investigations Report 2004–5137, 34 p.

Izbicki, J.A., Danskin, W.R., and Mendez, G.O., 1998, Chemistry and isotopic composition of ground water along a section near the Newmark area, San Bernardino County, California: U.S. Geological Survey Water-Resources Investigations Report 97–4179, 27 p.

Johnson, T.D., and Belitz, K., 2009, Assigning land use to supply wells for the statistical characterization of regional groundwater quality—Correlating urban land use and VOC occurrence: Journal of Hydrology, v. 370, p. 100–108, doi:10.1016/j.jhydrol.2009.02.056.

Jurgens, B.C., McMahon, P.B., Chapelle, F.H., and Eberts, S.M., 2009, An Excel® workbook for identifying redox processes in ground water: U.S. Geological Survey Open-File Report 2009–1004, 8 p. Available at http://pubs.usgs.gov/of/2009/1004/.

Kent, Robert, and Belitz, Kenneth, 2009, Ground-water quality data in the Upper Santa Ana Watershed Study Unit, November 2006–March 2007—Results from the California GAMA Program: U.S. Geological Survey Data Series 404, 116 p. Available at http://pubs.usgs.gov/ds/404.

Kent, Robert, Belitz, Kenneth, Altmann, A.J., Wright, M.T., and Mendez, G.O., 2005, Occurrence and distribution of pesticide compounds in surface water of the Santa Ana Basin, California, 1998–2001: U.S. Geological Survey Scientific Investigations Report 2005–5203, 143 p. Available at http://pubs.usgs.gov/sir/2005/5203/.

Kulongoski, J., and Belitz, K., 2004, Groundwater ambient monitoring and assessment program: U.S. Geological Survey Fact Sheet 2004–3088, 2 p.

Kulongoski, J.T., Belitz, Kenneth, Landon, M.K., and Farrar, Christopher, 2010, Status and understanding of groundwater quality in the North San Francisco Bay groundwater basins, 2004—California GAMA Priority Basin Project: U.S. Geological Survey Scientific Investigations Report 2010–5089, 88 p.

Landon, M.K., Belitz, Kenneth, Jurgens, B.C., Kulongoski, J.T., and Johnson, T.D., 2010, Status and understanding of groundwater quality in the Central-Eastside San Joaquin Basin, 2006—California GAMA Priority Basin Project: U.S. Geological Survey Scientific Investigations Report 2009–5266, 97 p.

Lindburg, R.D., and Runnells, D.D., 1984, Groundwater redox reactions: Science, v. 225, p. 925–927.

Lucas, L.L., and Unterweger, M.P., 2000, Comprehensive review and critical evaluation of the half-life of tritium: Journal of Research of the National Institute of Standards and Technology, v. 105, no. 4, p. 541–549.

Manning, A.H., Solomon, D.K., and Thiros, S.A., 2005, ^{3}H/^{3}He age data in assessing the susceptibility of wells to contamination: Ground Water, v. 43, no. 3, p. 353–367.

McMahon, P.B., and Chapelle, F.H., 2008, Redox processes and water quality of selected principal aquifer systems: Ground Water, v. 46, no. 2, p. 259–271.

Michel, R., and Schroeder, R., 1994, Use of long-term tritium records from the Colorado River to determine timescales for hydrologic processes associated with irrigation in the Imperial Valley, California: Applied Geochemistry, v. 9, p. 387–401.

Michel, R.L., 1989, Tritium deposition in the continental United States, 1953–83: U.S. Geological Survey Water-Resources Investigations Report 89–4072, 46 p.

Miller, C.L., Burton, S., and Manning, K., 2007, Preserving the Chino Basin: Civil Engineering, v. 77, no. 5, p. 56–61 and 81–82.

Nakagaki, N., Price, C.V., Falcone, J.A., Hitt, K.J., and Ruddy, B.C., 2007, Enhanced National Land Cover Data 1992 (NLCDe 92): U.S. Geological Survey Raster digital data, accessed January 11, 2012, at http://water.usgs.gov/lookup/getspatial?nlcde92.

Nakagaki, N., and Wolock, D.M., 2005, Estimation of agricultural pesticide use in drainage basins using land cover maps and county pesticide data: U.S. Geological Survey Open-File Report 2005–1188, 46 p.

Oki, D.S., and Giambelluca, T.W., 1987, DBCP, EDB and TCP contamination of groundwater in Hawaii: Ground Water, v. 25, no. 6, p. 693–702.

Orris, G.J., Harvey, G.J., Tsui, D.T., and Eldrige, J.E., 2003, Preliminary analyses for perchlorate in selected natural materials and their derivative products: U.S. Geological Survey Open-File Report 03–314, 6 p.

Piper, A.M., 1944, A graphic procedure in the geochemical interpretation of water analyses: American Geophysical Union Transactions, v. 25, p. 914–923.

Plummer, L.N., Michel, R.L., Thurman, E.M., and Glynn, P.D., 1993, Environmental tracers for age-dating young ground water, in Alley, W.M., ed., Regional groundwater quality, New York, Van Nostrand Reinhold, p. 255–294.

Rausser, G.C., Adams, G.D., Montgomery, W.D., and Smith, A.E., 2004, The social costs of an MTBE ban in California: Giannini Foundation Research Report 349, 62 p., accessed May 17, 2010, at http://giannini.ucop.edu/ResearchReports/349_MTBE.pdf.

Rowe, B.L., Toccalino, P.L., Moran, M.J., Zogorski, J.S., and Price, C.V., 2007, Occurrence and potential human-health relevance of volatile organic compounds in drinking water from domestic wells in the United States: Environmental Health Perspectives, v. 115, no. 11, p. 1539–1546.

Scott, J.C., 1990, Computerized stratified random site selection approaches for design of a groundwater quality sampling network: U.S. Geological Survey Water-Resources Investigations Report 90–4101, 109 p.

Solomon, D.K., and Cook, P.G., 2000, ^{3}H and ^{3}He, in Cook, P.G., and Herczeg, A.L., eds., Environmental tracers in subsurface hydrology: Boston, Kluwer Academic Press, p. 397–424.

State of California, 1999, Supplemental Report of the 1999 Budget Act 1999–00 Fiscal Year, Item 3940-001-0001, State Water Resources Control Board, accessed September 19, 2011, at http://www.lao.ca.gov/1999/99-00_supp_rpt_lang.html#3940.

State of California, 2001a, Assembly Bill No. 599, Chapter 522, accessed September 19, 2011, at http://www.swrcb.ca.gov/gama/docs/ab_599_bill_20011005_chaptered.pdf.

State of California, 2001b, Groundwater Monitoring Act of 2001: California Water Code, part 2.76, Section 10780-10782.3, accessed September 19, 2011, at http://www.leginfo.ca.gov/cgi-bin/displaycode?section=wat&group=10001-11000&file=10780-10782.3.

Tilman, F.D., 2009, Results of the analyses for 1,4-dioxane of groundwater samples collected in the Tucson Airport Remediation Project area, south-central Arizona, 2006–2009: U.S. Geological Survey Open-File Report 2009–1196, 14 p.

Toccalino, P.L., and Norman, J.E., 2006, Health-based screening levels to evaluate U.S. Geological Survey groundwater quality data: Risk Analysis, v. 26, no. 5, p. 1339–1348.

Toccalino, P.L., Norman, J.E., Phillips, R.H., Kauffman, L.J., Stackelberg, P.E., Nowell, L.H., Krietzman, S.J., and Post, G.B., 2004, Application of health-based screening levels to ground-water quality data in a state-scale pilot effort: U.S. Geological Survey Scientific Investigations Report 2004–5174, 14 p.

Tolstikhin, I.N., and Kamenskiy, I.L., 1969, Determination of groundwater ages by the T-^{3}He method: Geochemistry International, v. 6, p. 810–811.

Torgersen, T., Clarke, W.B., and Jenkins, W.J., 1979, The tritium/helium3 method in hydrology: IAEA-SM-228, v. 49, p. 917–930.

Urbansky, E.T., Brown, S.K., Magnuson, M.L., and Delty, C.A., 2001, Perchlorate levels in samples of sodium nitrate fertilizer derived from Chilean caliche: Environmental Pollution, v. 112, no. 3, p. 299–302.

U.S. Environmental Protection Agency, 1998, Code of Federal Regulations, title 40—protection of environment, chapter 1—environmental protection agency, subchapter E—pesticide programs, part 159—statements of policies and interpretations, subpart D—reporting requirements for risk/benefit information, 40 CFR 159.184: National Archives and Records Administration, September 19, 1997; amended June 19, 1998, accessed September 5, 2008, at http://www.epa.gov/EPA-PEST/1997/September/Day-19/p24937.htm.

U.S. Environmental Protection Agency, 1999, National primary drinking water regulations, radon-222: Federal Register, v. 64, no. 211, p. 59245–59294.

U.S. Environmental Protection Agency, 2000, National primary drinking water regulations; radionuclides; final rule, accessed February 2012 at https://federalregister.gov/a/00-30421.

U.S. Environmental Protection Agency, 2006, 2006 Edition of the Drinking Water Standards and Health Advisories, updated August 2006: U.S. Environmental Protection Agency, Office of Water EPA/822/R-06-013, http://www.epa.gov/waterscience/criteria/drinking/dwstandards.pdf.

U.S. Geological Survey, 2010, What is the Priority Basin Project?: U.S. Geological Survey website, accessed January 25, 2012, at http://ca.water.usgs.gov/gama.

Vogel, J.C., and Ehhalt, D., 1963, The use of the carbon isotopes in groundwater studies, *in* Radioisotopes in hydrology: Tokyo, IAEA, p. 383–395.

Wildermuth Environmental, Inc., 2000, TIN/TDS Phase 2A: Tasks 1 through 5, TIN/TDS study of the Santa Ana Watershed, Technical Memorandum, July 2000.

Woolfenden, L.R., and Kadhim, S., 1997, Geohydrology and water chemistry in the Rialto-Colton Basin, San Bernardino County California: U.S. Geological Survey Water-Resources Investigations Report 97-4012, 101 p.

Zebarth, B.J., Szeto, S.Y., Hii, B., Liebscher, H., and Grove, G., 1998, Groundwater contamination by chlorinated hydrocarbon impurities present in soil fumigant formulation: Water Quality Research Journal of Canada, v. 33, no. 1, p. 31–50.

Zogorski, J.S., Carter, J.M., Ivahnenko, T., Lapham, W.W., Moran, M.J., Rowe, B.L., Squillace, P.J., and Toccalino, P.L., 2006, Volatile organic compounds in the Nation's ground water and drinking-water supply wells: U.S. Geological Survey Circular 1292, 101 p.

Appendix A. Use of Data from the California Department of Public Health (CDPH) Database

California requires regular sampling of public-supply wells under Title 22. Historical data derived from these samples are available from the CDPH database. Assembly Bill 599 directs the Groundwater Ambient Monitoring and Assessment (GAMA) Program to use existing data and collect new data as needed. The GAMA Priority Basin Project uses these monitoring data along with newly collected data to characterize the water quality of the primary aquifers. The CDPH database provided additional water-quality data for the grid-based and spatially weighted approaches to estimating aquifer-scale proportions for a wide range of constituents. CDPH data were not used to supplement USGS-grid-well data for VOCs, pesticides, or perchlorate because reporting levels for these constituents in the CDPH database generally were not low enough to differentiate between "low" and "moderate" relative-concentrations.

Of the 107 grid cells in the study unit (including the 4 "equivalent cells" in the Elsinore Basin), 90 cells had USGS-grid data for organic constituents. Of these 90 cells, 49 also had USGS-grid data for inorganic constituents. Of the 107 grid cells, 17 did not have USGS-grid data for any constituents because no well was sampled (two of these—cell 1 for both the Cucamonga/Chino and the Yucaipa/San Timoteo study areas—did have inorganic data from wells sampled by the CDPH) (table A1). Three approaches were used to select CDPH inorganic constituent data for each grid cell where the USGS did not sample for inorganic constituents. The first step was to select CDPH data for the USGS-grid well (well sampled by the USGS for organic constituents, but not for inorganic constituents), provided that the CDPH data met quality-control criteria to minimize the selection of poor-quality data. Cation-anion balance was used as the quality-control assessment metric for selecting chemical analyses for a CDPH-grid well. Because water is electrically neutral, the total positive charge on dissolved cation species in a water sample must equal the total negative charge on dissolved anion species. The cation/anion imbalance commonly is used as a quality-control check for water sample analysis (Hem, 1992). Cation-anion balance was calculated as the difference between the total cations and total anions divided by the sum, expressed as a percentage:

$$\text{Percent difference} = \left(\frac{\left| \sum cations - \sum anions \right|}{\sum cations + \sum anions} \right) * 100,$$

where

$\sum cations$ is the sum of the concentrations of calcium, magnesium, sodium, and potassium in milliequivalents per liter (meq/L), and

$\sum anions$ is the sum of the concentrations of chloride, sulfate, fluoride, nitrate and bicarbonate in meq/L.

An imbalance, or percent difference, of greater than or equal to 10 percent indicates uncertainty in the quality of the data. The most recent CDPH data from USGS-grid wells were evaluated to determine whether the CDPH data had a cation/anion imbalance of less than 10 percent. If so, the CDPH inorganic data from the well were selected for use as grid well data for inorganic constituents. It was assumed that analyses with high-quality major ion data also had high-quality data for trace elements, nutrients, and radiochemical constituents. This step resulted in the selection of inorganic data from CDPH at 28 wells that were also USGS-grid wells. For identification purposes, data from the CDPH for these grid wells were assigned GAMA identification numbers equivalent to the GAMA USGS-grid well but with DG inserted between the study area prefix and sequence number (for example, CDPH-grid well USAWB-DG-03 is the same well as USGS-grid well USAWB-03, table A1).

If the first step did not yield CDPH inorganic data for a grid cell, the second step was to search the CDPH database to identify the highest randomly ranked well within that cell with a cation/anion imbalance of less than 10 percent. This step resulted in the selection of CDPH-grid wells for five grid cells, with CDPH inorganic data from a well not sampled by USGS as the grid well for that cell. These five CDPH-grid wells did not coincide with their cell's respective USGS-grid well. To identify these new CDPH-grid wells, a well ID was created that added DPH after the study area prefix and then added the grid-cell number for the study area (for example, CDPH-grid well USAWB-DPH-17).

If no CDPH well had data with a cation/anion balance of less than 10 percent, the third step was to select the highest ranked well in the CDPH database that had any of the needed inorganic data. This step resulted in selection of one USGS-grid well where CDPH data were used for nitrate plus nitrite only. Because the well was a USGS-grid well, a well ID was created that added DG to the GAMA ID (USAWB-DG-09).

The result of these steps was one grid well per cell having data from either the USGS database or the CDPH database, or a combination of data from both sources. Inorganic data were collected from 49 of the 90 USGS-grid wells (fig. A1). Inorganic data from 34 CDPH-grid wells in the CDPH database were used to supplement these data (fig. A2). Nitrate plus nitrite values were available for all 34 of these wells (table 2). As a result of combining data from

these 34 CDPH-grid wells with USGS-grid well inorganic data (49 wells), at least some inorganic data were available for 83 of the 107 grid cells.

Estimates of aquifer-scale proportion for constituents made on the basis of a smaller number of wells have a larger error associated with the 90 percent confidence intervals (based on the Jeffreys interval for the binomial distribution, Brown and others, 2001). Analysis of the combined datasets to evaluate the occurrence of high or moderate relative-concentrations for inorganic constituents was not affected by differences in reporting levels between USGS-grid and CDPH-grid data because concentrations above one-half of water-quality benchmarks (relative-concentration > 0.5) were generally substantially higher than the highest reporting levels. Comparisons between USGS-grid and CDPH-grid data are described in appendix B.

Table A1. Grid cell number, USGS-grid wells sampled, and CDPH-grid well identification number if CDPH data were used to supplement grid cell inorganic data, land-use categories, well construction information, and the relative elevation as the normalized position along a flowpath for wells sampled November 2006–March 2007 for the Upper Santa Ana Watershed study unit, California GAMA Priority Basin Project Upper Santa Ana Watershed study unit.

[CDPH, California Department of Public Health; USAWB, Bunker Hill/Cajon/Rialto-Colton study area well; USAWC, Cucamonga/Chino study area well; USAWE, Elsinore study area well; USAWR, Riverside-Arlington/Temescal study area well; USAWS, San Jacinto study area well; USAWY, Yucaipa/San Timoteo study area well; ft, feet; m, meter; LSD, land surface datum; na, not available; ns, cell not sampled (no data available); USGS, U.S. Geological Survey; USAWB-##, naming convention for USGS-grid well; USAWB-DG-##, naming convention for USGS understanding well; USAWB-DG-##, naming convention for CDPH data from a USGS-grid well; USAWB-DPH-##, naming convention for CDPH-data from a CDPH-grid well; PSW, public-supply well; DOM, domestic well; IND, industrial well; IRR, irrigation well; DES, desalter well; MON, monitoring well; REC, recreation use well (to fill ponds)]

Grid cell number	USGS GAMA well identification number	CDPH-grid well identification number	Well type	Agricultural land use within 500 m of the well (percent)	Natural land use within 500 m of the well (percent)	Urban land use within 500 m of the well (percent)	Well depth (ft below LSD)	Top of perforations (ft below LSD)	Bottom of perforations (ft below LSD)	Length from top of uppermost perforated interval to bottom of perforations (ft below LSD)	Relative elevation as the normalized position along flowpath (dimensionless) (value in parentheses is for separate well with CDPH data used to supplement GAMA data)
Bunker Hill/Cajon/Rialto-Colton study area grid wells											
1	USAWB-14	USGS data	PSW	0	22	78	1,060	300	1,040	740	0.627
2	USAWB-02	USGS data	PSW	0	96	4	234	61	230	169	0.924
3	USAWB-03	USAWB-DG-03	PSW	0	94	6	195	45	175	130	1.231
4	USAWB-09	USAWB-DG-09 [1]	PSW	0	79	21	126	52	126	74	1.749
5	ns	ns	ns	ns	ns	ns	ns	ns	ns	ns	ns
6	USAWB-05	USAWB-DG-05	PSW	0	94	6	50	21	48	27	1.050
7	USAWB-06	USAWB-DG-06	PSW	0	63	37	400	177	400	223	0.772
8	USAWB-04	USGS data	PSW	0	72	28	900	464	884	420	0.629
9	USAWB-08	USGS data	PSW	0	22	78	580	126	560	434	0.471
10	USAWB-18	USGS data	PSW	1	13	86	534	244	534	290	0.413
11	USAWB-01	USAWB-DG-01	PSW	0	45	55	1,067	497	1,047	550	0.395
12	USAWB-07	USGS data	PSW	0	36	64	654	240	462	222	0.347
13	USAWB-16	USGS data	PSW	0	22	78	963	888	951	63	0.366
14	USAWB-10	USGS data	PSW	0	0	100	708	490	680	190	0.437
15	USAWB-11	USGS data	PSW	0	5	95	606	378	569	191	0.512
16	USAWB-17	USAWB-DG-17	PSW	0	5	95	421	173	415	242	0.449
17	USAWB-12	USAWB-DPH-17	PSW	6	65	29	950	500	930	430	0.569 (0.494)
18	USAWB-15	USAWB-DG-15	PSW	16	65	19	1,020	600	1,000	400	0.563
19	USAWB-19	USAWB-DG-19	IRR	18	20	62	na	na	na	na	0.861
20	USAWB-13	USAWB-DG-13	PSW	4	17	79	474	150	442	292	0.556
Cucamonga/Chino study area grid wells											
1	no USGS well	USAWC-DPH-1	PSW	na	na	na	na	na	na	na	0.086
2	USAWC-23	USGS data	PSW	52	20	28	1,180	440	1,180	740	0.096
3	USAWC-05	USAWC-DG-05	PSW	1	17	82	800	300	775	475	0.222
4	USAWC-04	USGS data	PSW	0	18	82	560	317	535	218	0.249
5	USAWC-06	USAWC-DG-06	PSW	0	0	100	903	472	849	377	0.349
6	USAWC-16	USAWC-DG-16	PSW	17	16	67	491	230	460	230	0.915
7	USAWC-18	USAWC-DG-18	PSW	0	10	90	764	322	702	380	0.651

Table A1. Grid cell number, USGS-grid wells sampled, and CDPH-grid well identification number if CDPH data were used to supplement grid cell inorganic data, land-use categories, well construction information, and the relative elevation as the normalized position along a flowpath for wells sampled November 2006–March 2007 for the Upper Santa Ana Watershed study unit, California GAMA Priority Basin Project Upper Santa Ana Watershed study unit.—Continued

[CDPH, California Department of Public Health; USAWB, Bunker Hill/Cajon/Rialto-Colton study area well; USAWC, Cucamonga/Chino study area well; USAWE, Elsinore study area well; USAWR, Riverside-Arlington/Temescal study area well; USAWS, San Jacinto study area well; USAWY, Yucaipa/San Timoteo study area well; ft, feet; m, meter; LSD, land surface datum; na, not available; ns, cell not sampled (no data available); USGS, U.S. Geological Survey; USAWB-##, naming convention for USGS-grid well; USAWU-##, naming convention for USGS understanding well; USAWB-DG-##, naming convention for CDPH data from a USGS-grid well; USAWB-DPH-##, naming convention for CDPH-data from a CDPH-grid well; PSW, public supply well; DOM, domestic well; IND, industrial well; IRR, irrigation well; DES, desalter well; MON, monitoring well; REC, recreation use well (to fill ponds)]

Grid cell number	USGS GAMA well identification number	CDPH-grid well identification number	Well type	Land-use categories			Construction information				Relative elevation as the normalized position along flowpath (dimensionless) (value in parentheses is for separate well with CDPH data used to supplement GAMA data)
				Agricultural land use within 500 m of the well (percent)	Natural land use within 500 m of the well (percent)	Urban land use within 500 m of the well (percent)	Well depth (ft below LSD)	Top of perforations (ft below LSD)	Bottom of perforations (ft below LSD)	Length from top of uppermost perforated interval to bottom of perforations (ft below LSD)	
Cucamonga/Chino study area grid wells—Continued											
8	USAWC-14	USGS data	PSW	0	10	90	1,028	415	1,007	592	0.317
9	USAWC-24	USGS data	MON	21	38	41	300	280	300	20	0.237
10	USAWC-10	USAWC-DPH-10	PSW	18	39	43	520	290	500	210	0.090 (0.103)
11	USAWC-09	USAWC-DG-09	PSW	0	69	31	na	na	na	na	0.075
12	USAWC-12	USGS data	PSW	80	14	6	na	na	na	na	0.096
13	USAWC-25	USGS data	IRR	26	36	38	435	230	430	200	0.191
14	USAWC-15	USAWC-DG-15	PSW	64	13	23	1,028	415	1,007	592	0.281
15	USAWC-17	USGS data	PSW	26	19	55	900	370	885	515	0.333
16	USAWC-21	USGS data	PSW	12	5	83	1,110	400	1,090	690	0.772
17	USAWC-19	USAWC-DG-19	PSW	0	100	0	Spring	Spring	Spring	na	2.148
18	ns	ns	ns	ns	ns	ns	ns	ns	ns	ns	ns
19	USAWC-13	USAWC-DG-13	PSW	36	23	40	na	na	na	na	0.322
20	USAWC-08	USGS data	PSW	21	32	47	370	210	370	160	0.197
21	USAWC-11	USGS data	PSW	13	18	69	350	200	350	150	0.173
22	USAWC-07	USAWC-DG-07	PSW	0	25	75	218	113	213	100	0.121
23	USAWC-22	no CDPH data	PSW	0	10	90	[2]7,31	na	na	na	0.361
24	USAWC-01	USGS data	PSW	1	20	78	795	na	na	na	0.397
25	USAWC-02	USGS data	PSW	3	17	80	950	na	na	na	0.461
26	USAWC-20	USGS data	PSW	4	20	76	918	na	na	na	0.512
27	USAWC-03	USAWC-DG-03	PSW	0	10	90	1,040	500	1,030	530	0.623
Elsinore study area grid wells											
1	USAWE-01	USGS data	PSW	20	65	15	1,430	420	1,410	990	0.525
3	USAWE-02	USGS data	PSW	0	98	2	1,720	380	1,700	1,320	0.512
3	USAWE-03	USGS data	PSW	20	13	67	980	570	980	410	0.556
4	USAWE-04	USGS data	PSW	21	55	24	740	300	730	430	0.518

Table A1. Grid cell number, USGS-grid wells sampled, and CDPH-grid well identification number if CDPH data were used to supplement grid cell inorganic data, land-use categories, well construction information, and the relative elevation as the normalized position along a flowpath for wells sampled November 2006–March 2007 for the Upper Santa Ana Watershed study unit, California GAMA Priority Basin Project Upper Santa Ana Watershed study unit.—Continued

[CDPH, California Department of Public Health; USAWB, Bunker Hill/Cajon/Rialto-Colton study area well; USAWC, Cucamonga/Chino study area well; USAWE, Elsinore study area well; USAWR, Riverside-Arlington/Temescal study area well; USAWS, San Jacinto study area well; USAWY, Yucaipa/San Timoteo study area well; ft, feet; m, meter; LSD, land surface datum; na, not available; ns, cell not sampled (no data available); USGS, U.S. Geological Survey; USAWB-##, naming convention for USGS-grid well; USAWU-##, naming convention for USGS understanding well; USAWB-DG-##, naming convention for CDPH data from a USGS-grid well; USAWB-DPH-##, naming convention for CDPH-data from a CDPH-grid well; PSW, public supply well; DOM, domestic well; IND, industrial well; IRR, irrigation well; DES, desalter well; MON, monitoring well; REC, recreation use well (to fill ponds)]

Grid cell number	USGS GAMA well identification number	CDPH-grid well identification number	Well type	Land-use categories			Construction information				Relative elevation as the normalized position along flowpath (dimensionless) (value in parentheses is for separate well with CDPH data used to supplement GAMA data)
				Agricultural land use within 500 m of the well (percent)	Natural land use within 500 m of the well (percent)	Urban land use within 500 m of the well (percent)	Well depth (ft below LSD)	Top of perforations (ft below LSD)	Bottom of perforations (ft below LSD)	Length from top of uppermost perforated interval to bottom of perforations (ft below LSD)	
Riverside/Arlington-Temescal study area grid wells											
1	ns	ns	ns	ns	ns	ns	ns	ns	ns	ns	ns
2	USAWR-12	USGS data	PSW	0	6	94	500	200	250	50	0.163
3	USAWR-10	USGS data	PSW	0	13	87	220	108	204	96	0.102
4	USAWR-03	USGS data	PSW	49	20	32	183	24	170	146	0.183
5	USAWR-06	USGS data	IRR	0	17	83	190	120	180	60	0.214
6	USAWR-07	USGS data	DES	8	11	81	170	80	150	70	0.149
7	USAWR-04	USAWR-DPH-7	PSW	0	70	30	89	28	89	61	0.208 (0.120)
8	USAWR-09	USGS data	IRR	0	17	83	346	60	346	286	0.222
9	USAWR-01	USAWR-DG-01	PSW	33	27	40	na	na	na	na	0.262
10	USAWR-11	USGS data	PSW	1	57	42	644	250	634	384	0.407
11	USAWR-05	USGS data	DOM	13	46	41	192	na	na	na	0.254
12	USAWR-08	USGS data	PSW	0	28	72	402	40	380	340	0.275
13	USAWR-02	USAWR-DG-02	PSW	5	67	28	613	284	600	316	0.224
San Jacinto study area grid wells											
1	ns	ns	ns	ns	ns	ns	ns	ns	ns	ns	ns
2	USAWS-12	USGS data	PSW	7	73	20	760	240	740	500	0.184
3	USAWS-04	no CDPH data	IND	0	91	9	150	50	150	100	0.256
4	USAWS-01	USAWS-DG-01	PSW	6	22	72	225	100	205	105	0.369
5	USAWS-08	USGS data	PSW	16	9	76	428	170	420	250	0.602
6	ns	ns	ns	ns	ns	ns	ns	ns	ns	ns	ns
7	USAWS-03	no CDPH data	IRR	72	26	2	160	55	140	85	0.207
8	USAWS-09	USAWS-DG-09	DES	68	31	1	370	140	350	210	0.162
9	USAWS-10	no CDPH data	DES	16	75	9	360	90	320	230	0.090
10	USAWS-11	no CDPH data	DES	61	30	9	380	80	340	260	0.114
11	ns	ns	ns	ns	ns	ns	ns	ns	ns	ns	ns
12	USAWS-05	USGS data	PSW	26	15	59	625	100	620	520	0.129
13	USAWS-13	no CDPH data	DES	74	19	8	410	230	390	160	0.105
14	ns	ns	ns	ns	ns	ns	ns	ns	ns	ns	ns
15	ns	ns	ns	ns	ns	ns	ns	ns	ns	ns	ns

Table A1. Grid cell number, USGS-grid wells sampled, and CDPH-grid well identification number if CDPH data were used to supplement grid cell inorganic data, land-use categories, well construction information, and the relative elevation as the normalized position along a flowpath for wells sampled November 2006–March 2007 for the Upper Santa Ana Watershed study unit, California GAMA Priority Basin Project Upper Santa Ana Watershed study unit.—Continued

[CDPH, California Department of Public Health; USAWB, Bunker Hill/Cajon/Rialto-Colton study area well; USAWC, Cucamonga/Chino study area well; USAWE, Elsinore study area well; USAWR, Riverside-Arlington/Temescal study area well; USAWS, San Jacinto study area well; USAWY, Yucaipa/San Timoteo study area well; ft, feet; m, meter; LSD, land surface datum; na, not available; ns, cell not sampled (no data available); USGS, U.S. Geological Survey; USAWB-##, naming convention for USGS-grid well; USAWU-##, naming convention for USGS understanding well; USAWB-DG-##, naming convention for CDPH data from a USGS-grid well; USAWB-DPH-##, naming convention for CDPH-data from a CDPH-grid well; PSW, public supply well; DOM, domestic well; IND, industrial well; IRR, irrigation well; DES, desalter well; MON, monitoring well; REC, recreation use well (to fill ponds)]

Grid cell number	USGS GAMA well identification number	CDPH-grid well identification number	Well type	Land-use categories			Construction information				Relative elevation as the normalized position along flowpath (dimensionless) (value in parentheses is for separate well with CDPH data used to supplement GAMA data)
				Agricultural land use within 500 m of the well (percent)	Natural land use within 500 m of the well (percent)	Urban land use within 500 m of the well (percent)	Well depth (ft below LSD)	Top of perforations (ft below LSD)	Bottom of perforations (ft below LSD)	Length from top of uppermost perforated interval to bottom of perforations (ft below LSD)	
San Jacinto study area grid wells—Continued											
16	ns	ns	ns	ns	ns	ns	ns	ns	ns	ns	ns
17	ns	ns	ns	ns	ns	ns	ns	ns	ns	ns	ns
18	USAWS-16	USGS data	PSW	50	41	9	na	na	na	na	0.156
19	USAWS-02	USGS data	PSW	48	22	30	518	104	518	414	0.208
20	ns	ns	ns	ns	ns	ns	ns	ns	ns	ns	ns
21	USAWS-07	no CDPH data	REC	55	27	18	300	na	na	na	0.119
22	USAWS-15	USGS data	PSW	5	4	91	328	na	na	na	0.423
23	ns	ns	ns	ns	ns	ns	ns	na	na	ns	ns
24	USAWS-17	no CDPH data	IRR	51	45	4	na	na	na	na	0.254
25	USAWS-19	no CDPH data	IRR	76	22	3	na	na	na	na	0.184
26	USAWS-21	USGS data	IRR	67	32	1	900	440	900	460	0.223
27	USAWS-06	USGS data	PSW	43	14	43	650	200	650	450	0.336
28	USAWS-14	USGS data	PSW	55	13	32	1,550	470	1,530	1,060	0.471
29	USAWS-18	USGS data	PSW	0	24	75	714	324	714	390	0.473
30	USAWS-20	USGS data	PSW	12	15	73	696	300	676	376	0.412
31	ns	ns	ns	ns	ns	ns	ns	ns	ns	ns	ns
Yucaipa/San Timoteo study area grid wells											
1	no USGS well	USAWY-DPH-1 [1]	PSW	na	na	na	na	na	na	na	0.981
2	ns	ns	ns	ns	ns	ns	ns	ns	ns	ns	ns
3	USAWY-03	USAWY-DG-03	PSW	19	17	64	600	150	576	426	1.084
4	USAWY-04	USGS data	PSW	0	10	90	1,710	705	1,690	985	1.077
5	USAWY-02	USAWY-DG-02	PSW	8	44	48	596	290	584	294	1.423
6	USAWY-05	USGS data	PSW	0	17	83	585	320	585	265	1.322
7	USAWY-06	USGS data	PSW	11	45	45	790	150	340	190	1.321
8	USAWY-01	USAWY-DG-01	PSW	4	75	21	350	300	315	15	1.357
9	ns	ns	ns	ns	ns	ns	ns	ns	ns	ns	ns
10	USAWY-09	USAWY-DG-09	PSW	2	4	94	946	320	694	374	1.452
11	USAWY-08	USAWY-DG-08	PSW	9	33	58	444	58	434	376	1.784
12	USAWY-07	USAWY-DG-07	PSW	17	63	20	314	164	314	150	1.618

Table A1. Grid cell number, USGS-grid wells sampled, and CDPH-grid well identification number if CDPH data were used to supplement grid cell inorganic data, land-use categories, well construction information, and the relative elevation as the normalized position along a flowpath for wells sampled November 2006–March 2007 for the Upper Santa Ana Watershed study unit, California GAMA Priority Basin Project Upper Santa Ana Watershed study unit.—Continued

[CDPH, California Department of Public Health; USAWB, Bunker Hill/Cajon/Rialto-Colton study area well; USAWC, Cucamonga/Chino study area well; USAWE, Elsinore study area well; USAWR, Riverside-Arlington/Temescal study area well; USAWS, San Jacinto study area well; USAWY, Yucaipa/San Timoteo study area well; ft, feet; m, meter; LSD, land surface datum; na, not available; ns, cell not sampled (no data available); USGS, U.S. Geological Survey; USAWB-##, naming convention for USGS-grid well; USAWU-##, naming convention for USGS understanding well; USAWB-DG-##, naming convention for CDPH data from a USGS-grid well; USAWB-DPH-##, naming convention for CDPH-data from a CDPH-grid well; PSW, public supply well; DOM, domestic well; IND, industrial well; IRR, irrigation well; DES, desalter well; MON, monitoring well; REC, recreation use well (to fill ponds)]

Grid cell number	USGS GAMA well identification number	CDPH-grid well identification number	Well type	Land-use categories			Construction information				Relative elevation as the normalized position along flowpath (dimensionless) (value in parentheses is for separate well with CDPH data used to supplement GAMA data)
				Agricultural land use within 500 m of the well (percent)	Natural land use within 500 m of the well (percent)	Urban land use within 500 m of the well (percent)	Well depth (ft below LSD)	Top of perforations (ft below LSD)	Bottom of perforations (ft below LSD)	Length from top of uppermost perforated interval to bottom of perforations (ft below LSD)	
USGS-understanding wells											
RAT-09	USAWU-01	USGS data	PSW	0	32	68	170	50	170	120	0.198
YUTI-04	USAWU-02	USGS data	PSW	10	17	72	1,100	390	1,100	710	1.091
off grid	USAWU-03	no inorganic data	PSW	0	46	54	220	180	220	40	2.973
RAT-10	USAWU-04	USGS data	PSW	1	23	75	540	320	520	200	0.386
BUNCO-09	USAWU-05	USGS data	PSW	0	1	99	na	310	787	na	0.542
CUCHI-04	USAWU-06	USGS data	PSW	0	24	76	800	295	784	489	0.280
CUCHI-20	USAWU-07	USGS data	PSW	39	25	36	980	400	980	580	0.304
CUCHI-02	USAWU-08	USGS data	PSW[3]	53	20	27	na	na	na	na	0.097
CUCHI-14	USAWU-09	USGS data	MON	7	10	83	310	29	310	281	0.211

[1] Inorganic data limited to nitrate plus nitrite from CDPH (no USGS inorganic data).

[2] Well depth determined from hole depth reported on driller's log.

[3] Water from this well was not being served to the public at the time of this study.

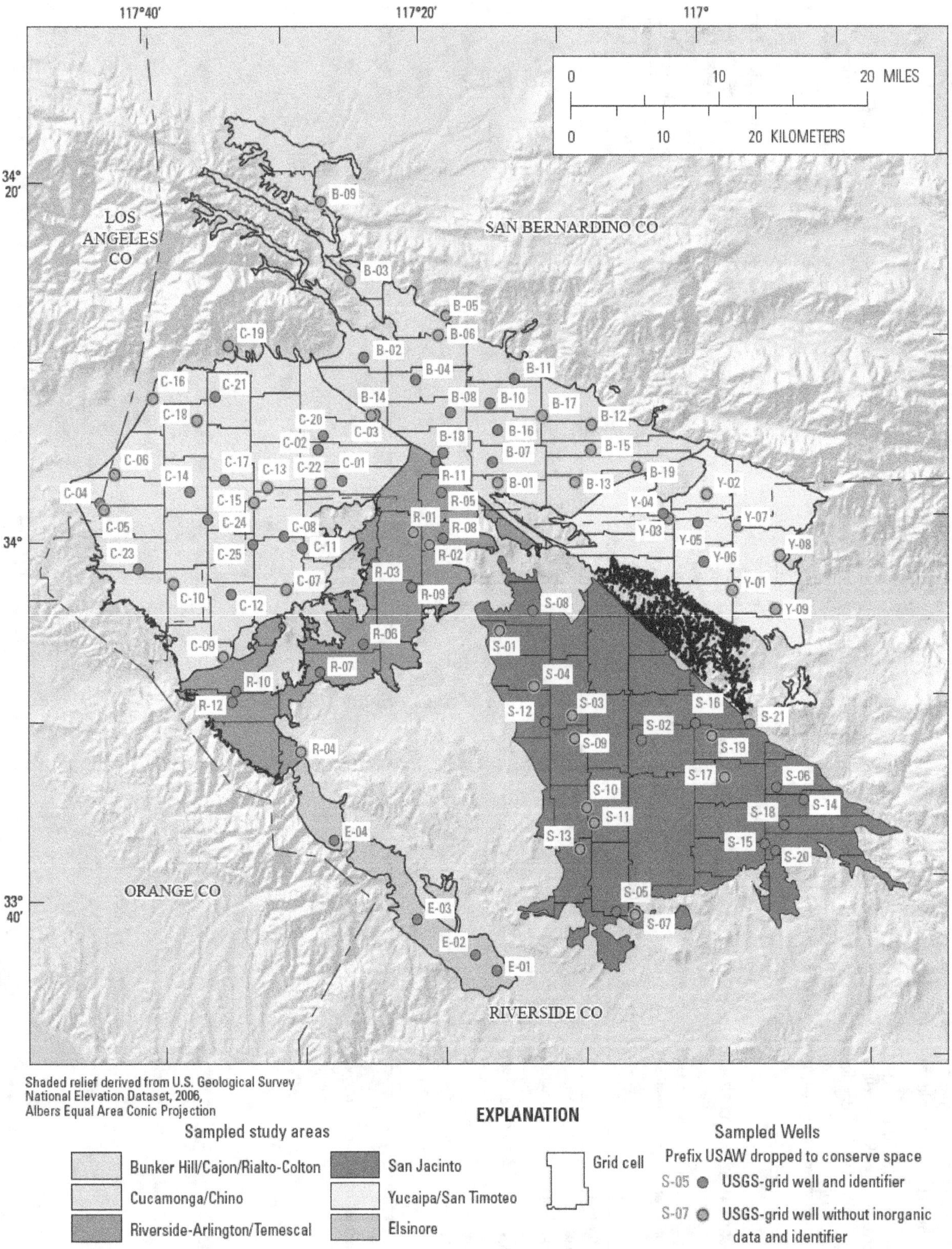

Shaded relief derived from U.S. Geological Survey
National Elevation Dataset, 2006,
Albers Equal Area Conic Projection

EXPLANATION

Sampled study areas

- Bunker Hill/Cajon/Rialto-Colton
- Cucamonga/Chino
- Riverside-Arlington/Temescal
- San Jacinto
- Yucaipa/San Timoteo
- Elsinore

Grid cell

Sampled Wells

Prefix USAW dropped to conserve space

S-05 ● USGS-grid well and identifier

S-07 ◉ USGS-grid well without inorganic data and identifier

Figure A1.   Identifiers and locations of USGS-grid wells sampled during November 2006–March 2007, Upper Santa Ana Watershed study unit, GAMA Priority Basin Project.

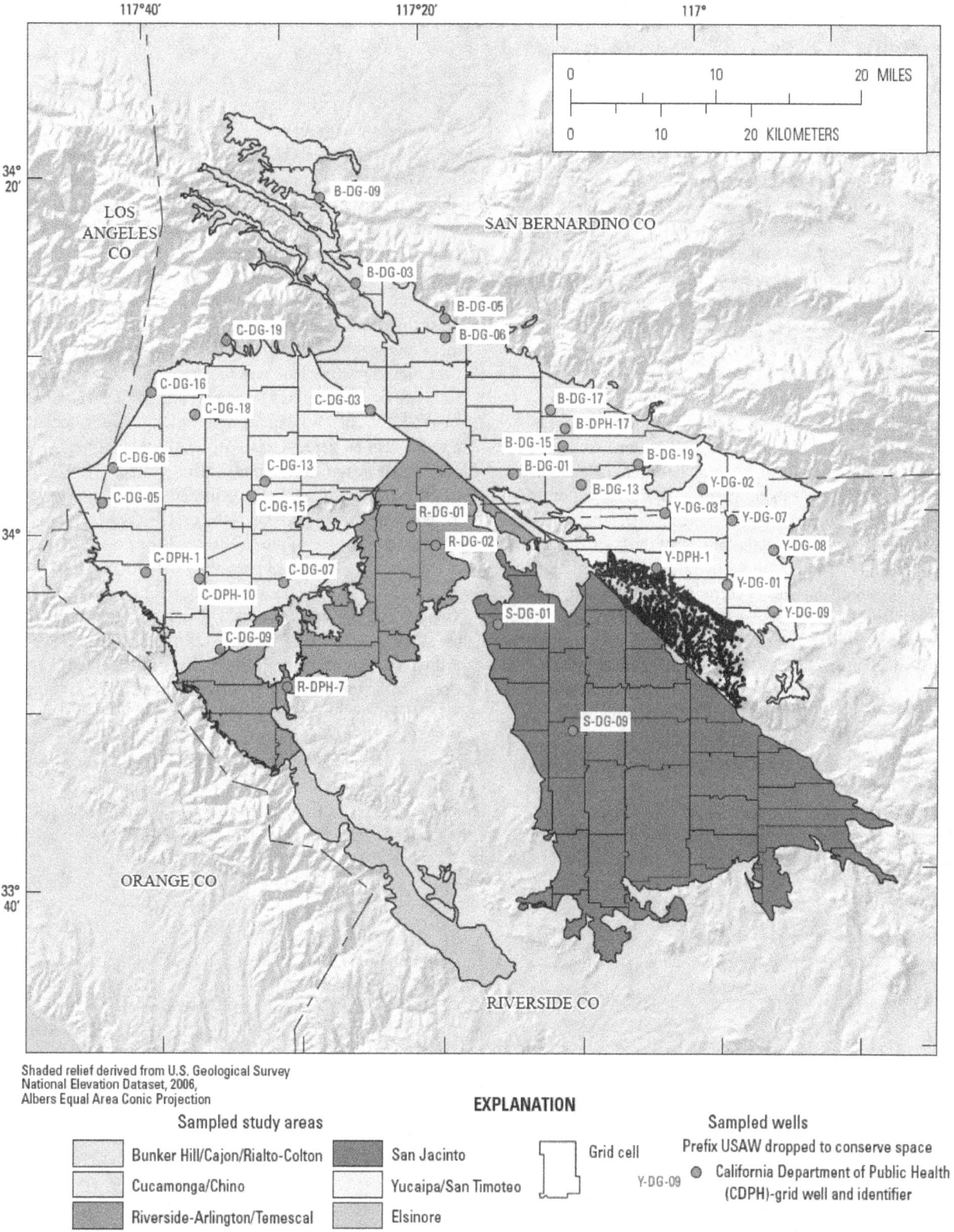

Figure A2.   Identifiers and locations of CDPH-grid wells from which data for inorganic constituents from the California Department of Public Health database were used to supplement USGS data, Upper Santa Ana Watershed study unit, GAMA Priority Basin Project.

Appendix B. Comparison of California Department of Public Health and U.S. Geological Survey-GAMA Data

CDPH and USGS-GAMA data were compared to assess the validity of using data from these different sources. Because USGS laboratory reporting levels (LRL) for most organic constituents and trace elements were substantially lower than the method detection limits (MDL) used to report CDPH data (table 3), it was generally not possible to meaningfully compare concentrations of these constituent types in individual wells. However, concentrations of major ions and nitrate, which generally are prevalent and have concentrations substantially above LRLs and MDLs, were compared for each well with data from both sources. Comparisons were made for wells that were analyzed by USGS-GAMA for inorganic constituents and had data within the most recent 3-year interval in the CDPH database. Forty-one wells had major ion and nitrate data in common between the datasets. Wilcoxon signed rank tests of paired analyses for ten of these constituents (fig. B1) indicated significant differences between USGS-GAMA and CDPH data for two of the constituents: calcium (p=0.032), and fluoride (p<0.001). While differences between the paired datasets occurred for these constituents, most sample pairs plotted close to a 1-to-1 line (fig. B1). The relative percent difference (RPD) was calculated for each data pair. The median RPD was 7.3 percent; 83 percent of the RPD values were less than 20 percent. These direct comparisons indicated that the GAMA and CDPH data for major ions and nitrate were not significantly different.

Major ion data for grid wells with the requisite analyses of cations (sodium, potassium, calcium, and magnesium) and anions (sulfate, chloride, carbonate, bicarbonate, nitrite, and nitrate) were plotted on Piper diagrams (Piper, 1944) with CDPH major ion data to determine whether the grid wells represented the range of groundwater types that have historically been observed in the study unit. Piper diagrams show the relative abundance of major cations and anions (on a charge equivalent basis) as a percentage of the total ion content of the water (fig. B2). Piper diagrams often are used to define groundwater type (Hem, 1992). All CDPH data from the period November 30, 2003, to December 1, 2006 (prior period) having cation/anion data and a cation/anion imbalance of less than 10 percent were retrieved and plotted on these Piper diagrams for comparison with grid well data.

Calcium bicarbonate was the dominant water type for the 57 USGS-GAMA wells (51 percent) and for the 477 CDPH wells (62 percent). When no single cation or anion accounts for more than 50 percent of its group, the water type is described as *mixed cation* or *mixed anion* (Hem, 1992). Twenty percent of the CDPH wells had a *mixed cation* water type compared to 33 percent of the USGS-GAMA wells. Similarly, 16 percent of the CDPH wells had a *mixed anion* water type compared to 23 percent of the USGS-GAMA wells. Sodium was the dominant cation for the relatively small proportion of wells (8 percent) with a dominant cation other than calcium or mixed. Chloride was the dominant anion for the relatively small proportion of wells (3 percent) with a dominant anion other than bicarbonate or mixed, although six CDPH wells (1 percent) had sulfate as their dominant anion. This similarity of the range of relative abundance of major cations and anions in USGS wells to the set of CDPH wells indicates that the USGS-grid and understanding wells represent most of the diversity of water types present within the Upper Santa Ana Watershed study unit.

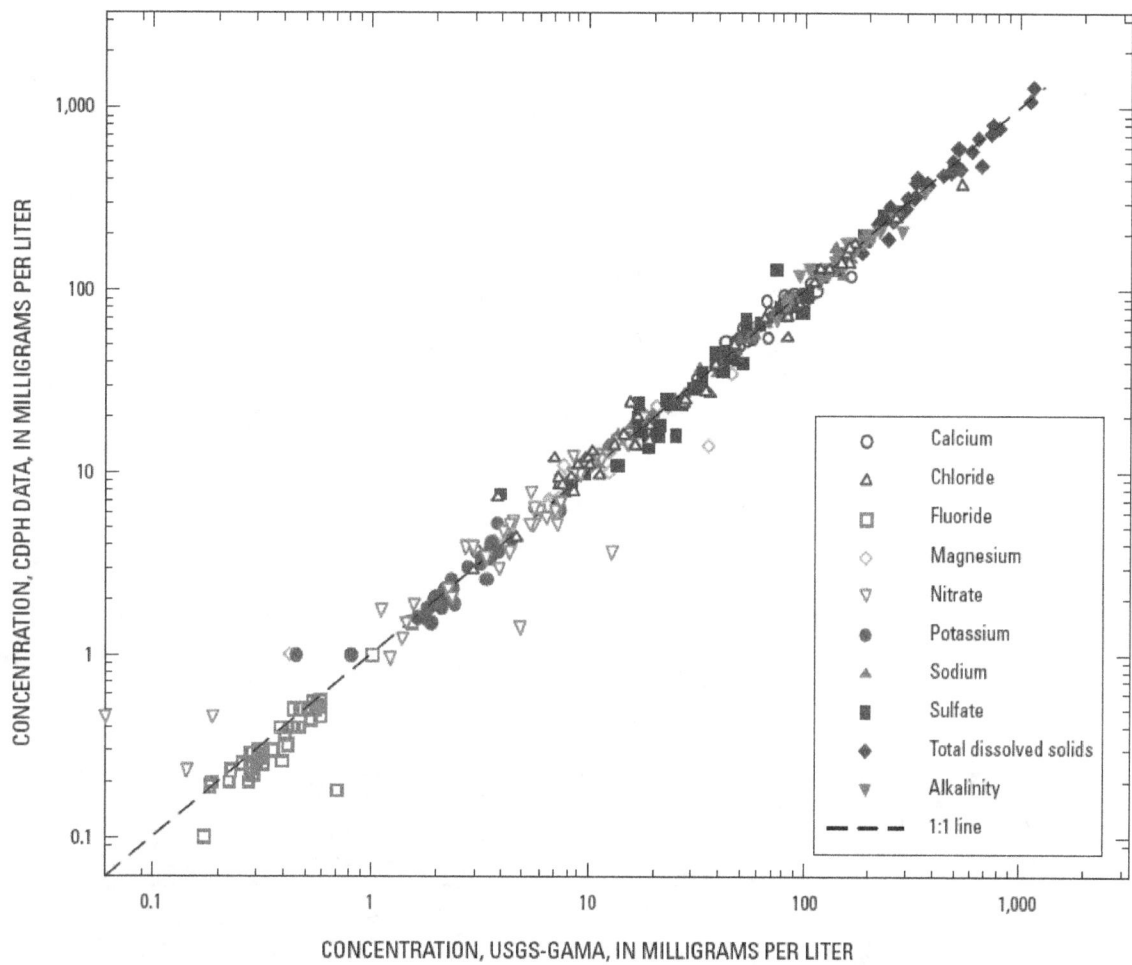

Figure B1. Paired inorganic concentrations from wells sampled by the Groundwater Ambient Monitoring and Assessment (GAMA) Program, November 2006–March 2007, and prior 3-year California Department of Public Health data, Upper Santa Ana Watershed study unit, GAMA Priority Basin Project.

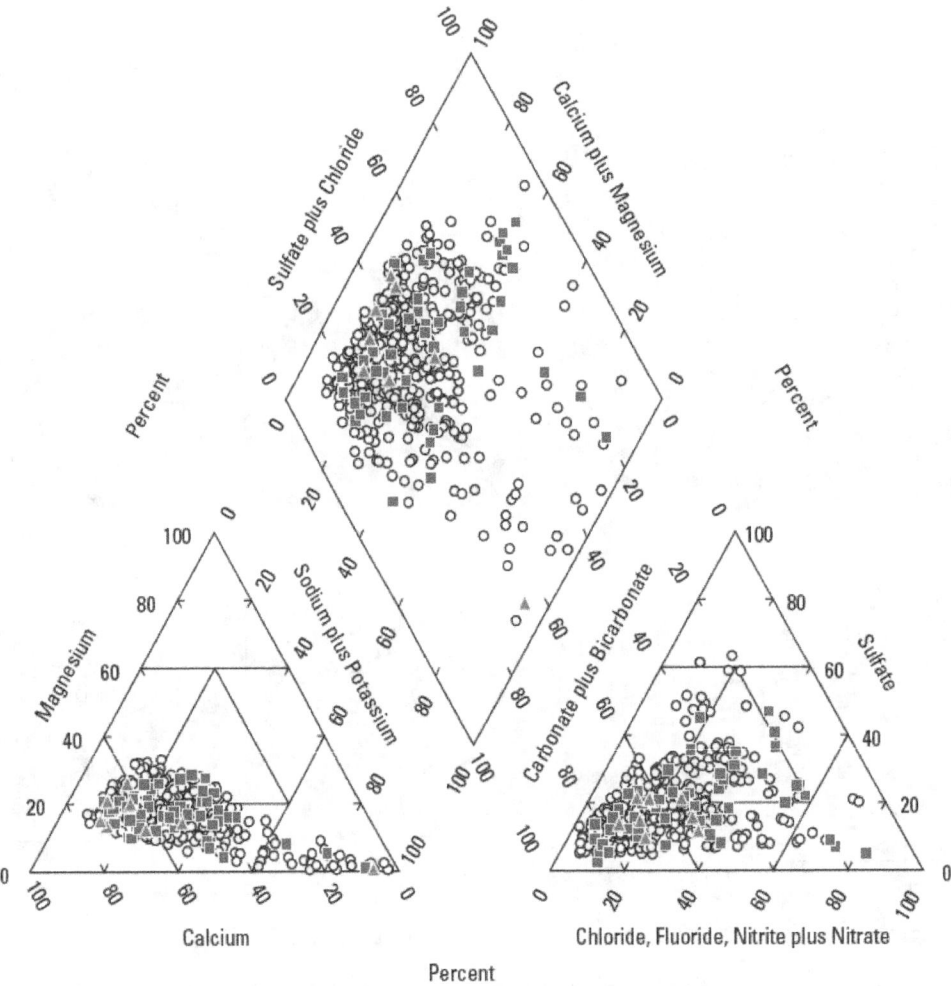

EXPLANATION

■ USGS-grid well
▲ USGS-understanding well
○ CDPH well

Figure B2.   Piper diagram for water types in USGS-GAMA wells and wells in the California Department of Public Health database, Upper Santa Ana Watershed study unit, GAMA Priority Basin Project.

Appendix C. Calculation of Aquifer-Scale Proportions

The *status assessment* is intended to characterize the quality of groundwater resources in the primary aquifers of the USAW study unit. The primary aquifers are defined by the perforated depth intervals of the wells listed in the CDPH database. The use of the term "primary aquifers" does not imply that there exists a discrete aquifer unit. In most groundwater basins, municipal and community supply wells generally are perforated at greater depths than domestic wells. Thus, because domestic wells are not listed in the CDPH database, the primary aquifers generally correspond to the portion of the aquifer system tapped by municipal and community supply wells. A majority of the wells used in the *status assessment* are listed in the CDPH database, and are therefore classified as municipal and community drinking-water supply wells. However, to the extent that domestic wells are perforated over the same depth intervals as the CDPH wells, the assessments presented in this report also may be applicable to the portions of the aquifer systems used for domestic drinking-water supplies.

Two statistical approaches, grid-based and spatially weighted, were selected to evaluate the aquifer-scale proportions of the area of the primary aquifers in the USAW study unit with high, moderate, or low relative-concentrations of constituents relative to water-quality benchmarks (Belitz and others, 2010). The grid-based and spatially weighted estimations of aquifer-scale proportions, based on a spatially distributed grid cell network across the USAW study unit, are intended to characterize the water quality of the primary aquifers, or at depths from which drinking water is usually drawn. These approaches assign weights to wells based on a single well per cell (grid-based) or the number of wells per cells (spatially weighted).

Raw detection frequencies, derived from the percentage of the total number of wells with high or moderate relative-concentrations, also were calculated for individual constituents, but were not used for estimating aquifer-scale proportion because this method creates spatial bias towards regions with large numbers of wells.

1. Grid-based. One well in each grid cell, a "grid well," was randomly selected to represent the area of the primary aquifers (Belitz and others, 2010). Most grid wells sampled for the USAW study unit were USGS-grid wells. However, data for all constituents were not available for some USGS-grid wells, so additional data for CDPH-grid wells were selected to provide data for grid cells with no USGS-grid wells. The relative-concentration for each constituent (concentration relative to its benchmark) was then evaluated for each grid well. The proportion of the primary aquifers (by area) with high relative-concentrations was calculated by dividing the number of cells with concentrations greater than the benchmark (relative-concentration greater than 1) by the total number of grid wells in the USAW study unit. Proportions containing moderate relative-concentrations were calculated similarly. Confidence intervals for grid-based aquifer-scale proportions were computed using the Jeffreys interval for the binomial distribution (Brown and others, 2001). The grid-based estimate is spatially unbiased because the cells represented are equal areas. However, the grid-based approach may not identify constituents that exist at high concentrations in small proportions of the primary aquifers.

2. Spatially weighted. The spatially weighted approach relied on USGS-grid well data collected from November 2006 to March 2007, and CDPH data from November 30, 2003, to December 1, 2006 (most recent analyses per well for all wells within each grid cell), and USGS-understanding public-supply well data. However, instead of data from only one well per grid cell, the spatially weighted approach uses all wells in each cell to calculate the high, moderate, and low relative-concentrations for the cell. The high, moderate, and low aquifer-scale proportions are then calculated from the percentage of cells with high, moderate, or low relative-concentrations (Isaaks and Srivastava, 1989). The resulting proportions are spatially unbiased (Isaaks and Srivastava, 1989), again, because the cells represented are equal areas. Confidence intervals for spatially weighted estimates of aquifer-scale proportion are not described in this report.

The raw detection frequency approach merely is the percentage (frequency) of wells within the USAW study unit with high relative-concentrations. It was calculated by considering all of the available data from November 30, 2003, to December 1, 2006, for the CDPH well data (the most recent analysis per well for all wells), the USGS-grid well data, and understanding wells. However, this approach is not spatially unbiased because the CDPH and understanding wells are not uniformly distributed. Consequently, high values (or low values) for wells clustered in a particular area represent a small part of the primary aquifers, and could be given a disproportionately high (or low) weight compared to that given by spatially unbiased approaches. Therefore, raw detection frequencies were not used to assess aquifer-scale proportions.

Appendix D. Ancillary Datasets

Land-Use Classification

Land use was classified by using an enhanced version of the satellite-derived [98-ft (30-m) pixel resolution] USGS National Land Cover Dataset (Nakagaki and others, 2007). This dataset has been used in previous national and regional studies relating land use to water quality (Gilliom and others, 2006; Zogorski and others, 2006). The dataset characterizes land cover during the early 1990s. One pixel in the dataset imagery represents a land area of 9,688 ft^2 (900 m^2), calculated from the pixel of 98 ft (30 m). The imagery was classified into 25 land-cover classifications (Nakagaki and Wolock, 2005). These 25 land-cover classifications were assigned to three general classifications for the purpose of general categorization of principal land use: urban, agricultural, and natural.

Land-use statistics for the study unit, study areas, and for circles with a radius of 1,640 ft (500 m) around each study well were calculated based on these classified datasets using the software ArcGIS. The 500-m buffer represents a contributing area as defined and evaluated by Johnson and Belitz (2009). These are given in table A1. Figure 5 displays the land-use map based on the calculation of land use in the study areas from aerial coverage (30-meter pixel = 900 square meters).

Average land use around grid wells (radius of 500 m) in the USAW study unit was dominated by urban land use (51 percent of the area) and natural landscape (33 percent of the area), while agricultural land use accounted for 17 percent of the area (fig. 4). Average land use across the study unit was more evenly distributed among these three major classifications, with 35 percent urban land use, 44 percent natural landscape, and 21 percent agricultural land use in the study unit. However, it appears that land use in the 500-m radius surrounding the sampled grid wells represented land use in the study areas reasonably well, with one exception (fig. D1A). Grid wells sampled in the Yucaipa/San Timoteo study area had, on average, land use surrounding them that was about 30 percent more urban, and about 20 percent less natural landscape than did the study area as a whole (figs. D1A, D4). The higher percentage of urbanized land surrounding the grid wells reflects the association of public-supply wells with population density. The area surrounding grid wells, particularly for the Yucaipa/San Timoteo study area, may reflect greater urban influence than might be expected on the basis of the average land use of the study areas. Land-use proportions for each grid and understanding well sampled by the USGS are shown in figure D1B.

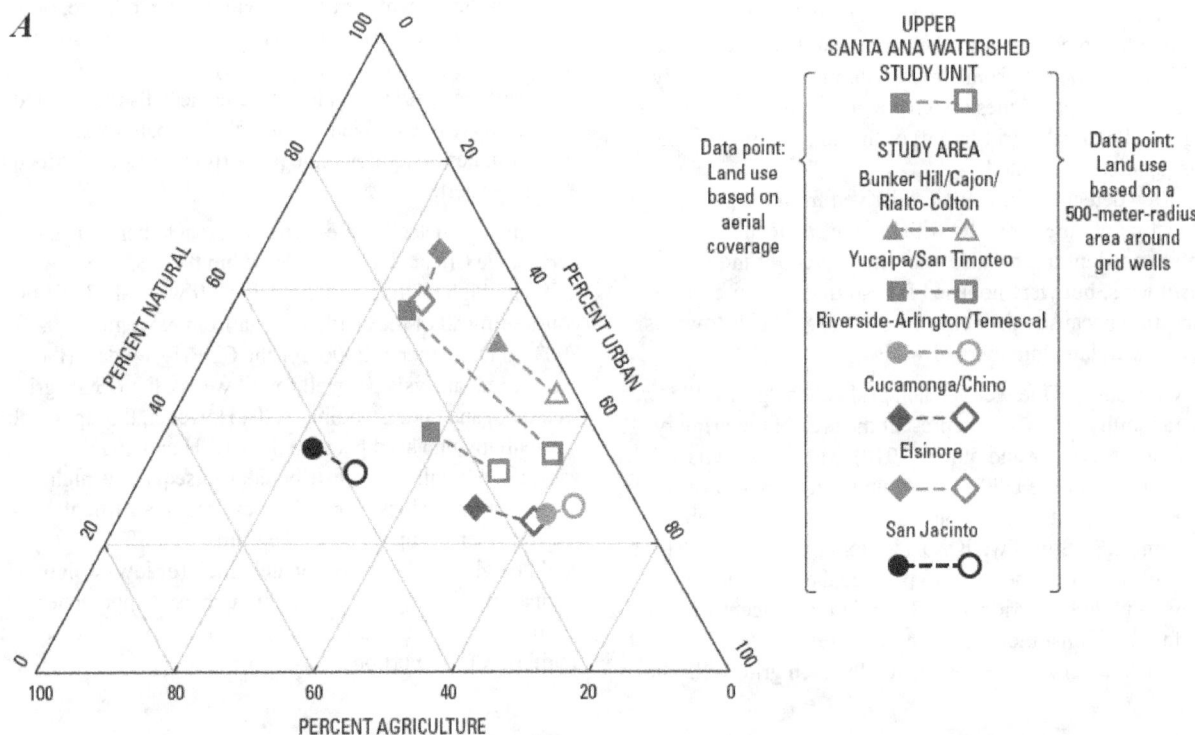

Figure D1A.   Ternary diagram showing proportions of urban, agricultural, and natural land-use categories in each study unit (solid symbols), and for the USGS wells sampled (open symbols) for the Upper Santa Ana Watershed study unit, GAMA Priority Basin Project. [Land uses were determined from USGS National Land Cover Data from Nakagaki and others (2007)]

B

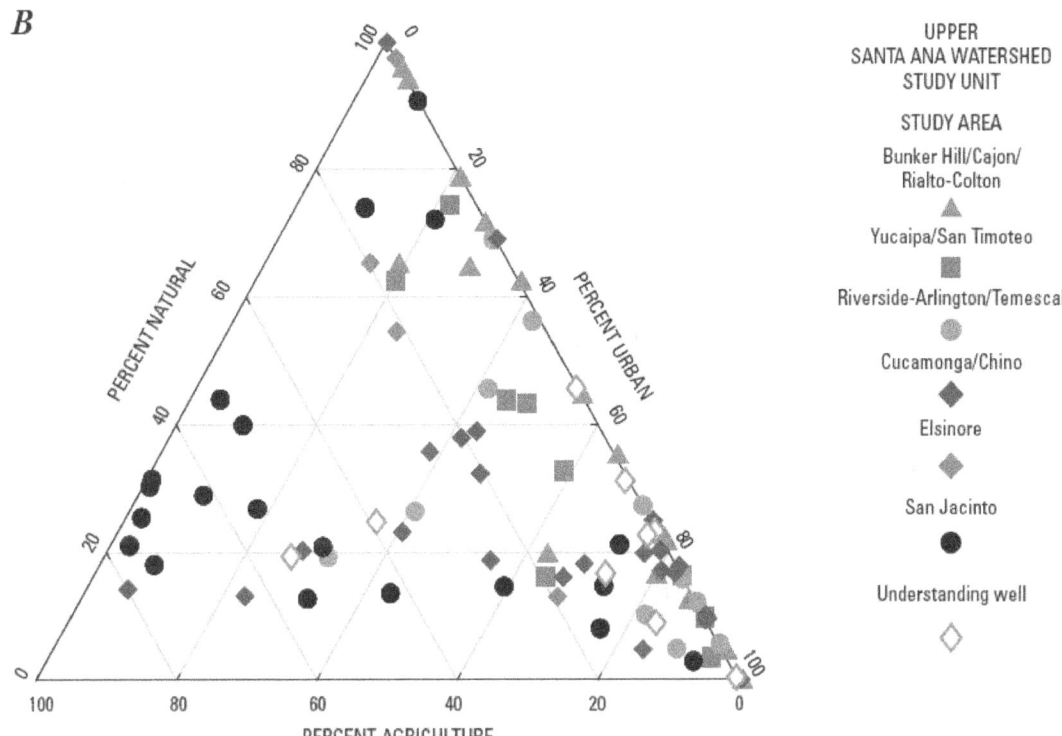

Figure D1*B*   Ternary diagram showing proportions of urban, agricultural, and natural land-use categories within a 500-meter radius surrounding grid and understanding wells sampled for the Upper Santa Ana Watershed study unit, GAMA Priority Basin Project. [Land uses were determined from USGS National Land Cover Data from Nakagaki and others (2007).]

Well Construction Information

Well construction data were derived, in part, from drillers' logs. Other sources of well construction data included ancillary records provided by well owners and the USGS National Water Information System database. Well identification verification procedures are described by Kent and Belitz (2009). Well depths and depths to the top- and bottom-of-perforations for USGS-grid, understanding, and CDPH-grid wells are listed in table A1. USAW wells were classified as public-supply wells, irrigation wells, desalter wells, monitoring wells, an industrial well, a "recreational" well, and a domestic well. Public-supply wells pump the groundwater from the aquifer to a distribution system. Irrigation wells supply water for agriculture and are generally located near fields where crops are grown. Desalter wells extract high-salinity groundwater for treatment and subsequent use by homes, industry, or agriculture. Monitoring wells tend to be short-screened wells installed exclusively for monitoring purposes. Domestic wells pump groundwater from the aquifer for home use. The grid well classified as "recreational" (USAWS-07) is used to maintain water hazards on a golf

course (table A1). Most USAW grid wells were production wells used for public supply. However, the seventeen grid wells that were not public-supply wells had screened intervals at depths similar to those of the public-supply grid wells.

Understanding wells were selected to better understand groundwater quality, including the movement of groundwater and changes in chemistry along approximate flow paths. Eight of the nine understanding wells were public-supply wells, and one (USAWU-09) was a monitoring well (table A1).

Depths of USGS- and CDPH-grid wells varied across the study unit. Grid wells had depths ranging from 50 to 1,720 ft (15 to 524 m) below land surface; the median was 580 ft (177 m) (fig. D2; table A1). Depths to the tops of the perforations ranged from 21 to 888 ft (6 to 271 m), with a median of 230 ft (70 m). The perforation length (distance from the top to the bottom perforation) was up to 1,320 ft (402 m) with a median of 287 ft (87 m). The understanding wells have ranges in well depth and depth to top of perforations very similar to those of the grid wells. Well construction information was available for 80 of the 90 grid wells and 8 of the 9 understanding wells sampled in the USAW study unit. (table A1).

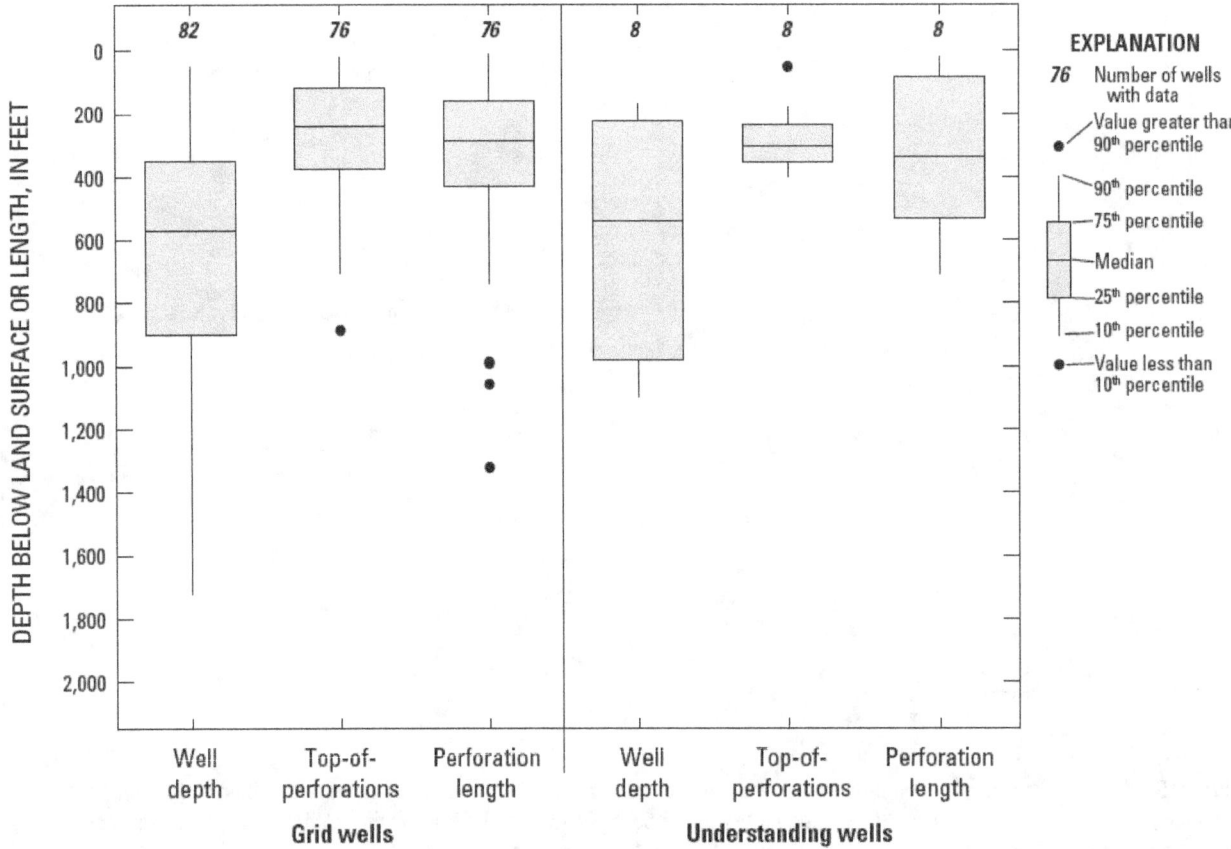

Figure D2.   Construction characteristics for grid and understanding wells, Upper Santa Ana Watershed study unit, California GAMA Priority Basin Project.

Groundwater Age Classification

Groundwater age data and classifications are listed in table D1. Groundwater dating techniques indicate the time since the groundwater was last in contact with the atmosphere. Techniques used to estimate groundwater residence times or "age" include those based on tritium (for example, Tolstikhin and Kamenskiy, 1969; Torgersen and others, 1979) and carbon-14 activities (^{14}C) (for example, Vogel and Ehhalt, 1963; Plummer and others, 1993).

Tritium (^3H) is a short-lived radioactive isotope of hydrogen with a half-life of 12.32 years (Lucas and Unterweger, 2000). Tritium is produced naturally in the atmosphere from the interaction of cosmogenic radiation with nitrogen (Craig and Lal, 1961), by above-ground nuclear explosions, and by the operation of nuclear reactors. Tritium enters the hydrologic cycle following oxidation to tritiated water. Tritium values in precipitation under natural conditions would be about 3 to 15 TU (Craig and Lal, 1961; Clark and Fritz, 1997). Above-ground nuclear explosions resulted in a large increase in tritium values in precipitation, beginning in about 1952 and peaking in 1963 at values over 1,000 TU in

the northern hemisphere (Michel, 1989). Radioactive decay over a period of 50 years would decrease tritium values of 10 TU to 0.6 TU.

Previous investigations have used a range of tritium values from 0.3 to 1.0 TU as thresholds for indicating presence of water that has exchanged with the atmosphere since 1952 (Michel, 1989; Plummer and others, 1993; Michel and Schroeder, 1994; Clark and Fritz, 1997; Manning and others, 2005). For samples collected for the USAW study unit in 2006–2007, tritium values greater than a threshold of 1.0 TU were defined as indicating presence of groundwater recharged since 1952. By using a tritium value of 1.0 TU for the threshold in this study, the age classification scheme allows a slightly larger fraction of modern groundwater to be classified as pre-modern than if a lower threshold were used. A lower threshold for tritium would result in fewer samples classified as pre-modern than mixed, when carbon-14 would suggest that they were primarily pre-modern. This higher threshold was considered more appropriate for this study because many of the wells were production wells with long screens and mixing of waters of different ages is likely to occur. Tritium activities of the water samples are listed in table D1.

Table D1.   Summary of groundwater age data and classification into modern, mixed, and pre-modern age categories for samples collected during November 2006 through March 2007, Upper Santa Ana Watershed study unit, California, GAMA Priority Basin Project.

[C degrees Celsius; TU, tritium units; ^{14}C, carbon–14; pmc, percent modern carbon; <, less than; blank field, no data]

GAMA_ID	Tritium activity (TU)	Uncorrected carbon-14 age (years)	Tritium uncertainty (TU)	[14]C (pmc)[1]	[14]C counting uncertainty (pmc)	Age classification
USAWB-01	0.00		0.12			Pre-modern
USAWB-02	3.41	<1,000	0.31	98	0.34	Modern
USAWB-03	2.72		0.14			Modern
USAWB-04	2.91	<1,000	0.31	91	0.29	Modern
USAWB-05	2.70		0.16			Modern
USAWB-06	2.92		0.15			Modern
USAWB-07	1.31	1050	0.18	87	0.3	Mixed
USAWB-08	2.69	<1,000	0.31	98	0.31	Modern
USAWB-09	2.65		0.13			Modern
USAWB-10	2.69	<1,000	0.31	101	0.44	Modern
USAWB-11	1.59	<1,000	0.18	96	0.31	Modern
USAWB-12	3.33		0.15			Modern
USAWB-13	3.16		0.16			Modern
USAWB-14	1.69	1,060	0.18	87	0.34	Mixed
USAWB-15	3.50		0.17			Modern
USAWB-16	0.00	1,940	0.35	78	0.32	Pre-modern
USAWB-17	2.43		0.32			Modern
USAWB-18	0.66	1,150	0.13	86	0.35	Pre-modern
USAWB-19	3.33		0.19			Modern
USAWC-01	0.04	1,300	0.15	84	0.35	Pre-modern
USAWC-02	0.50	1,540	0.16	82	0.34	Pre-modern
USAWC-03	1.73		0.16			Mixed
USAWC-04	1.50	<1,000	0.31	90	0.36	Modern
USAWC-05	0.50		0.11			Pre-modern
USAWC-06	3.66		0.20			Modern
USAWC-07	1.62		0.38			Mixed
USAWC-08	0.40	1,020	0.31	87	0.35	Pre-modern
USAWC-09	2.14		0.14			Modern
USAWC-10	0.28		0.10			Pre-modern
USAWC-11	0.68	<1,000	0.31	101	0.39	Mixed
USAWC-12	0.74	<1,000	0.13	110	0.42	Mixed
USAWC-13	0.55		0.11			Pre-modern
USAWC-14	0.31	<1,000	0.31	88	0.35	Pre-modern
USAWC-15	0.20		0.05			Pre-modern
USAWC-16	3.58		0.21			Modern
USAWC-17	1.00	<1,000	0.31	90	0.36	Modern
USAWC-18	0.85		0.12			Pre-modern
USAWC-19	2.87		0.15			Modern
USAWC-20	0.44	1,530	0.12	82	0.34	Pre-modern
USAWC-21	0.40	<1,000	0.31	88	0.35	Pre-modern
USAWC-22	0.39		0.12			Pre-modern
USAWC-23	0.09	<1,000	0.18	88	0.35	Pre-modern
USAWC-24	6.33	<1,000	0.29	96	0.38	Modern
USAWC-25	1.24	1,550	0.32	82	0.35	Mixed
USAWE-01	2.50	2,560	0.31	72	0.4	Mixed
USAWE-02	0.59		0.18			Pre-modern
USAWE-03	1.69	1,,180	0.18	84	0.44	Mixed
USAWE-04	2.69	<1,000	0.31	104	0.37	Modern
USAWR-01	0.00		0.78			Pre-modern
USAWR-02	0.41		0.04			Pre-modern
USAWR-03	1.00	<1,000	0.18	100	0.31	Modern

Table D1.   Summary of groundwater age data and classification into modern, mixed, and pre-modern age categories for samples collected during November 2006 through March 2007, Upper Santa Ana Watershed study unit, California, GAMA Priority Basin Project.—Continued

[C degrees Celsius; TU, tritium units; ^{14}C, carbon–14; pmc, percent modern carbon; <, less than; blank field, no data]

GAMA_ID	Tritium activity (TU)	Uncorrected carbon-14 age (years)	Tritium uncertainty (TU)	[14]C (pmc)[1]	[14]C counting uncertainty (pmc)	Age classification
USAWR-04	4.16		0.17			Modern
USAWR-05	1.97	<1,000	0.10	94	0.3	Modern
USAWR-06	3.06	<1,000	0.20	108	0.42	Modern
USAWR-07	3.38	<1,000	0.40	105	0.41	Modern
USAWR-08	2.10	<1,000	0.31	107	0.42	Modern
USAWR-09	0.96	<1,000	0.15	98	0.39	Mixed
USAWR-10	5.92	<1,000	0.30	94	0.37	Modern
USAWR-11	0.68	1,230	0.31	85	0.35	Pre-modern
USAWR-12	1.41	1,310	0.31	84	0.34	Mixed
USAWS-01	1.96		0.13			Modern
USAWS-02	0.68	1,200	0.31	85	0.35	Pre-modern
USAWS-03	5.11		0.22			Modern
USAWS-04	4.93		0.24			Modern
USAWS-05	2.10	1,370	0.18	84	0.34	Mixed
USAWS-06	0.00	4,810	0.31	54	0.25	Pre-modern
USAWS-07	1.52		0.33			Mixed
USAWS-08	1.00	<1,000	0.31	98	0.39	Modern
USAWS-09	4.57		0.22			Modern
USAWS-10	3.19		0.20			Modern
USAWS-11	1.73		0.13			Mixed
USAWS-12	0.90		0.31			Pre-modern
USAWS-13	2.06		0.19			Modern
USAWS-14	0.40	3,590	0.18	63	0.28	Pre-modern
USAWS-15	0.00	2,100	0.18	76	0.32	Pre-modern
USAWS-16	0.23	7,650	0.13	38	0.21	Pre-modern
USAWS-17	1.66		0.14			Mixed
USAWS-18	0.00	2,180	0.18	75	0.32	Pre-modern
USAWS-19	0.59		0.15			Pre-modern
USAWS-20	0.59	<1,000	0.18	89	0.36	Pre-modern
USAWS-21	0.83	6,200	0.08	46	0.24	Pre-modern
USAWY-01	0.10		0.09			Pre-modern
USAWY-02	2.16		0.12			Modern
USAWY-03	0.14		0.04			Pre-modern
USAWY-04	0.00	2,460	0.09	73	0.52	Pre-modern
USAWY-05	0.31	2,030	0.18	77	0.32	Pre-modern
USAWY-06	1.00	<1,000	0.31	92	0.37	Modern
USAWY-07	2.01		0.17			Modern
USAWY-08	1.36		0.15			Mixed
USAWY-09	0.00		0.10			Pre-modern
USAWU-01	2.83	<1,000	0.14	99	0.31	Modern
USAWU-02	0.31		0.05			Pre-modern
USAWU-03	3.89		0.18			Modern
USAWU-04	1.06	1,040	0.08	87	0.35	Mixed
USAWU-05	1.26	1,310	0.13	84	0.34	Mixed
USAWU-06	3.01	2,340	0.20	74	0.31	Mixed
USAWU-07	0.36	1,450	0.10	83	0.34	Pre-modern
USAWU-08	0.22	13,060	0.05	19	0.15	Pre-modern
USAWU-09	3.62	<1,000	0.20	116	0.44	Modern

[1] Carbon-14 (pmc) values here differ slightly from the values reported in table 13 of Kent and Belitz (2009). Here carbon-14 values were normalized to a standard carbon-13 of –25 per mil (VPDB) and reported as percent modern. These values were converted to non-normalized values using the carbon-13 of the sample and converted to pmc using the calculation procedure described in Plummer and others (2004).

Carbon-14 (^{14}C) is a widely used chronometer based on the radiocarbon content of organic and inorganic carbon. Dissolved inorganic carbon species, carbonic acid, bicarbonate, and carbonate typically are used for ^{14}C dating of groundwater. ^{14}C is formed in the atmosphere by the interaction of cosmic-ray neutrons with nitrogen and, to a lesser degree, with oxygen and carbon. ^{14}C is incorporated into carbon dioxide and mixed throughout the atmosphere. The carbon dioxide enters the hydrologic cycle because it dissolves in precipitation and surface water in contact with the atmosphere. ^{14}C activity in groundwater, expressed as percent modern carbon (pmc), reflects the time since groundwater was last exposed to the atmospheric ^{14}C source. ^{14}C has a half-life of 5,730 years and can be used to estimate groundwater ages ranging from 1,000 to approximately 30,000 years before present.

The ^{14}C age (residence time, presented in years) is calculated on the basis of the decrease in ^{14}C activity as a result of radioactive decay since groundwater recharge, relative to an assumed initial ^{14}C concentration (Clark and Fritz, 1997). An average initial ^{14}C activity of 100 pmc is assumed for this study, with estimated errors on calculated groundwater ages up to ± 20%. Calculated ^{14}C ages (table D1) in this study are referred to as "uncorrected" because they have not been adjusted to consider exchanges with sedimentary sources of carbon (Fontes and Garnier, 1979). Groundwater with a ^{14}C activity of >88 pmc is reported as having an age of <1,000 years; no attempt is made to refine ^{14}C ages <1,000 years. Measured values of percent modern carbon can be >100 pmc because the definition of the ^{14}C activity in "modern" carbon does not include the excess ^{14}C produced in the atmosphere by above-ground nuclear weapons testing. For the USAW study unit, ^{14}C activity <90 pmc was defined as indicative of presence of groundwater recharged before the modern era. The threshold value of 90 pmc was selected because all groundwater samples with tritium <1.0 TU also had ^{14}C <90 pmc. ^{14}C values in table D1 expressed as percent modern carbon differ slightly from the values reported in table 13 of Kent and Belitz (2009) because the values in table D1 were normalized to a standard carbon-13 (^{13}C) of −25 per mil (VPDB) and reported as percent modern carbon.

In this study, the age distributions of samples are classified as pre-modern, modern, and mixed (table D1). Groundwater with tritium activity less than 1 tritium unit and ^{14}C less than 90 pmc is designated as pre-modern, defined as having been recharged before 1952. Groundwater with tritium activity greater than 1 TU and ^{14}C greater than 90 pmc is designated as modern, defined as having been recharged after 1952. Samples with pre-modern and modern components are designated as mixed groundwater, which includes substantial fractions of old and young waters. In reality, pre-modern groundwater could contain small fractions of modern groundwater, and modern groundwater could contain small fractions of pre-modern groundwater. Tritium concentrations, uncorrected carbon-14 age, and sample age classifications are reported in table D1. Although more sophisticated lumped parameter models used for analyzing age distributions that incorporate mixing are available (for example, Cook and Böhlke, 2000), use of these alternative models to characterize age mixtures was beyond the scope of this report. Rather, classification into modern (recharged after 1952), mixed, and pre-modern (recharged before 1952) categories was sufficient to provide an appropriate and useful characterization for the purposes of examining groundwater quality.

Of the 99 grid and understanding wells sampled by the USGS-GAMA Priority Basin Project, groundwater ages were classified as modern for 42 wells, mixed for 19 wells (evidence of modern and pre-modern groundwater in the same sample), and pre-modern for 38 wells (table D1). The areal distribution of the age classifications of the wells is shown in figure D3.

Relative Elevation of Wells

The relative elevation of wells within the alluvial valleys was an additional factor examined for the understanding of water quality in the USAW study unit (table A1). Groundwater in alluvium moves under a natural hydraulic gradient that conforms in a general way to the surface topography (Faye, 1973). In the Upper Santa Ana Valley, groundwater movement generally follows the path of the Santa Ana River, which flows from the eastern edge of the valley westward and southward towards the Prado Wetlands and Dam (fig. 2). In the San Jacinto Basin, groundwater movement generally follows the path of the San Jacinto River, which exits the San Jacinto Mountains in the southeastern part of the valley and flows westward to Lake Elsinore (fig. 2). Relative elevations for wells sampled in the USAW study unit were determined separately for these two flow systems in the following way. First, the elevations at which the Santa Ana and the San Jacinto Rivers enter and exit their respective valleys were established as maximum and baseline (minimum) elevations for the Upper Santa Ana Valley and the San Jacinto groundwater basins. Elevation ranges for the two basins were defined as the difference between their maximum and baseline land-surface elevations. The appropriate baseline elevation was then subtracted from the elevation of each well, and this difference was divided by the elevation range for the basin where the well was located (wells in the Elsinore groundwater basin used the Upper Santa Ana Valley baseline and range).

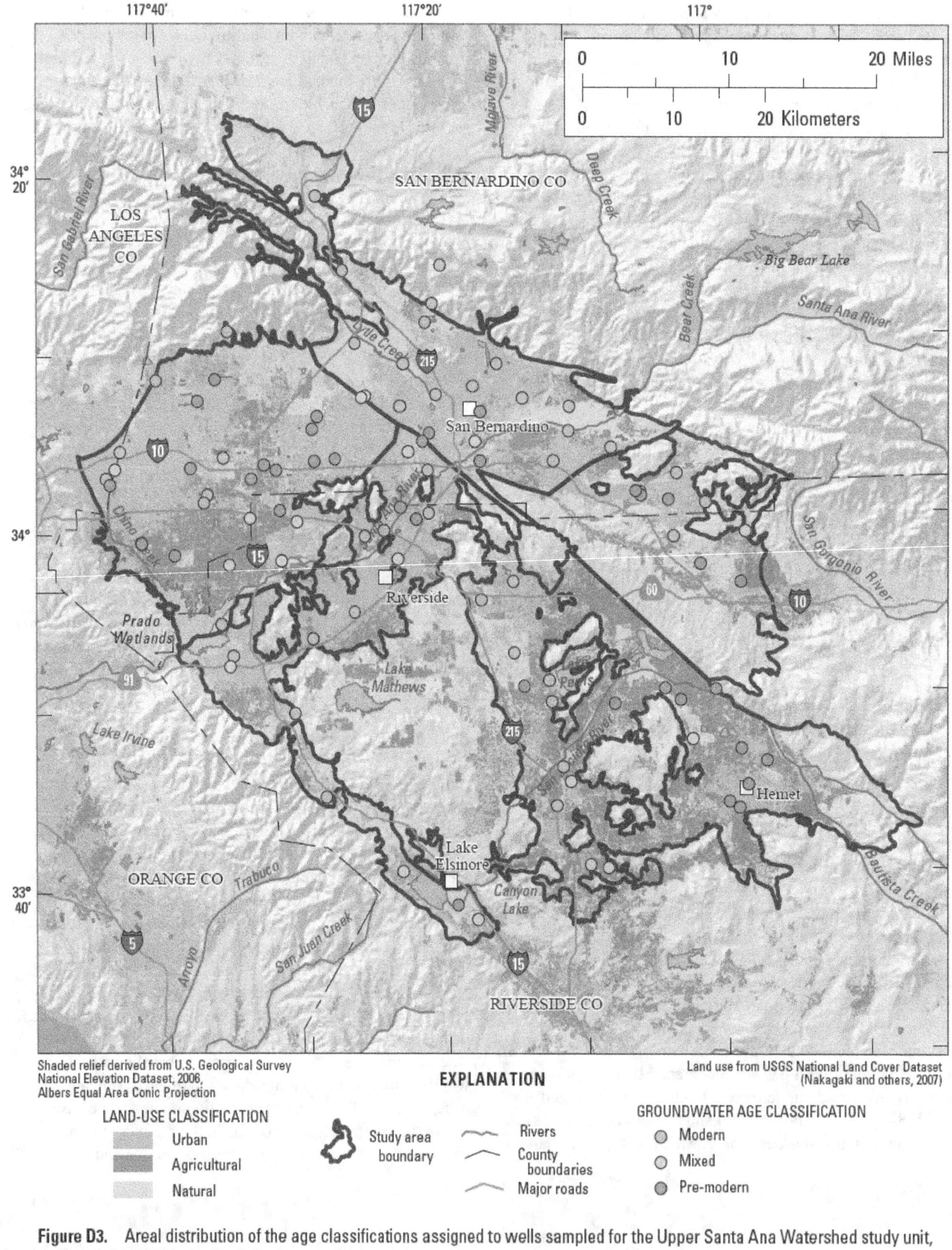
117°40′
117°20′
117°
34° 20′
34°
33° 40′
0 10 20 Miles
0 10 20 Kilometers
SAN BERNARDINO CO
LOS ANGELES CO
ORANGE CO
RIVERSIDE CO
Mojave River
Deep Creek
San Gabriel River
Big Bear Lake
Santa Ana River
Bear Creek
Lytle Creek
San Bernardino
Riverside
Chino Creek
Prado Wetlands
Lake Mathews
Lake Perris
San Jacinto River
Hemet
Lake Elsinore
Canyon Lake
Trabuco
San Juan Creek
Arroyo
Lake Irvine
San Gorgonio River
Bautista Creek
15
215
10
60
91
5
Shaded relief derived from U.S. Geological Survey National Elevation Dataset, 2006, Albers Equal Area Conic Projection
Land use from USGS National Land Cover Dataset (Nakagaki and others, 2007)
EXPLANATION
LAND-USE CLASSIFICATION
Urban
Agricultural
Natural
Study area boundary
Rivers
County boundaries
Major roads
GROUNDWATER AGE CLASSIFICATION
Modern
Mixed
Pre-modern

Figure D3. Areal distribution of the age classifications assigned to wells sampled for the Upper Santa Ana Watershed study unit, California GAMA Priority Basin Project.

The resulting relative elevation value is dimensionless and represents the relative position of the groundwater in each well along its flow path. A well that is at the same elevation as the river where it enters the valley would have a relative elevation of 1, while a well at the elevation of the river where it exits the valley would have a relative elevation of 0. Fifteen wells in the Upper Santa Ana Valley have relative elevation values that are greater than 1 because they are at elevations above that of the Santa Ana River where it enters the valley. Most (10) of these are in the Yucaipa/San Timoteo study area. Relative elevations are listed in table A1. Higher values of relative elevation indicate locations in the upgradient or proximal portion of the flow system, and lower values indicate locations in the downgradient or distal portion of the flow system.

Grid wells selected using a spatially distributed randomized design were distributed across the entire range of relative elevations in the study unit (table A1). However, there are differences in relative elevations among the study areas (fig. D4), which reflect the landscape positions of each of them. Wells in the San Jacinto study area are evaluated separately from the other study areas using baseline elevations for the San Jacinto River drainage instead of the Santa Ana River drainage. The two highest relative-elevation values are for wells which are located in the mountains adjacent to the study unit (USAWU-03 and USAWC-19, figs. 6, A1).

In general, the highest relative-elevation values were for grid wells in the Yucaipa/San Timoteo study area with a median of 1.34 (dimensionless). Next, the Bunker Hill/Cajon/Rialto-Colton study area grid wells had a median relative elevation of 0.57. Relative elevations for the four grid wells in the Elsinore study area varied little, from 0.51 to 0.56. Grid wells in the Cucamonga/Chino study area had a median relative elevation of 0.30. The lowest relative-elevation values were observed for grid wells sampled in the Riverside-Arlington/Temescal study area (where the Santa Ana River exits the study unit), with a median value of 0.22.

Grid wells in the San Jacinto study area are located at disproportionately low relative elevations. This likely reflects the dominant terrain of this groundwater basin, which has relatively high elevations only on the edges, along with some centrally-located hilly outcrops that were excluded from the study area. The median relative elevation for grid wells in the San Jacinto study area was 0.21.

Relative-elevation values for the understanding wells reflect those of the study areas where they are located. The highest value, for understanding wells as well as the entire study unit, is for USAWU-03, located in the San Bernardino mountains (figs. 6, D4). The relative elevation of another understanding well, USAWU-02, plots as an outlier for understanding wells (fig. D4), but its relative-elevation value is below the median for the Yucaipa/San Timoteo study area in which it is located. The other understanding wells are located in the Cucamonga/Chino study area (four wells), the Riverside-Arlington/Temescal study area (two wells), and the Bunker Hill/Cajon/Rialto-Colton study area (one well), and have relative-elevation values in the ranges of values for those study areas.

Geochemical Conditions

Geochemical conditions investigated as potential explanatory variables in this report include oxidation-reduction characteristics, dissolved oxygen (DO) concentrations, and pH (table D2). Oxidation-reduction (redox) conditions influence the mobility of many organic and inorganic constituents (McMahon and Chapelle, 2008). Along groundwater flow paths, redox conditions commonly proceed along a well-documented sequence of terminal electron acceptor processes (TEAPs); one TEAP typically is predominant at a particular time and aquifer location (Chapelle and others, 1995; Chapelle, 2001). The predominant TEAPs are oxygen-reduction (oxic), nitrate-reduction, manganese-reduction, iron-reduction, sulfate-reduction, and methanogenesis. The presence of redox-sensitive chemical species indicating more than one TEAP may indicate that (1) the well's discharge includes mixed waters from different redox zones upgradient of the well, (2) the well is screened across more than one redox zone, or (3) there is spatial heterogeneity in microbial activity in the aquifer. In addition, different redox couples often are not consistent, indicating electrochemical disequilibrium in groundwater (Lindburg and Runnells, 1984) complicating the assessments of redox conditions.

In this report, redox conditions were represented in two ways: as DO concentration and redox category on the basis of the predominant TEAP(s). DO concentrations were measured at USGS-grid and USGS-understanding wells (Kent and Belitz, 2009), but were not reported in the CDPH database (table D2). Redox conditions were classified on the basis of DO, nitrate, manganese, iron, and sulfate concentrations using the classification scheme of McMahon and Chapelle (2008) (table D2). An automated workbook program was used to assign the redox classification to each sample (Jurgens and others, 2009). For wells without USGS inorganic constituent data, the most recent data within the previous 3 years for that well in the CDPH database were used.

Groundwater in the USAW study unit is primarily oxic. Eighty-two of the 101 groundwater samples from wells having data for redox characterization are in the oxic category, where the predominant redox process is oxygen reduction. Samples from an additional three wells had similarly high DO concentrations, but also had dissolved iron concentrations greater than 100 micrograms per liter, suggesting mixed redox processes. A sample from an additional well had no DO data and a dissolved iron concentrations greater than 100 micrograms per liter, and this sample was also categorized as a mixed redox process. Samples from only eight wells had anoxic redox conditions. These anoxic wells were distributed across all USAW study areas except for the Bunker Hill/Cajon/Rialto-Colton and Yucaipa/San Timoteo study areas.

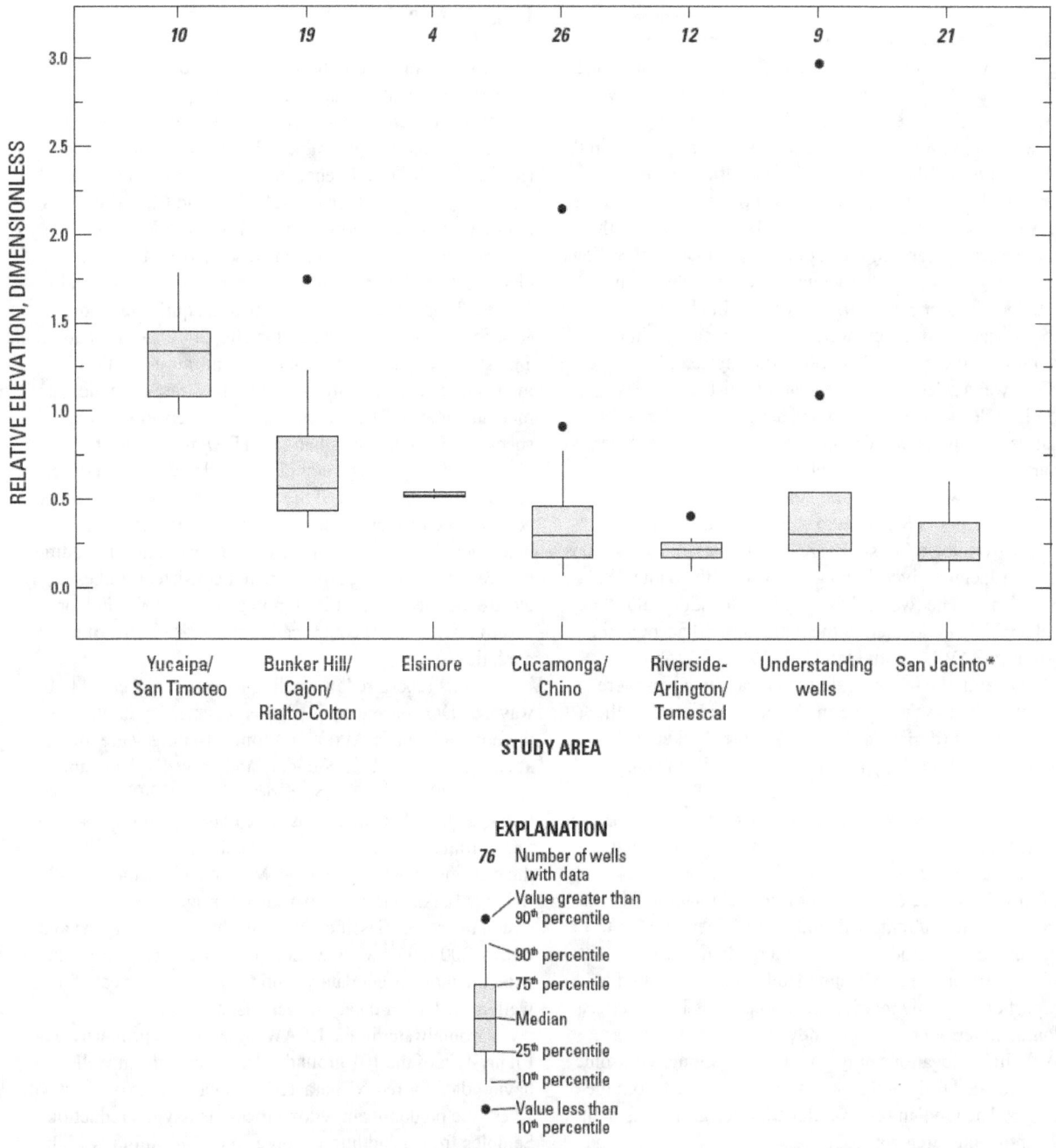

Figure D4.   Relative elevations for grid wells (by study area) and understanding wells, Upper Santa Ana Watershed study unit, California GAMA Priority Basin Project. *Wells in the San Jacinto study area were evaluated separately from the other study areas and understanding wells because their evaluation uses baseline elevations for the San Jacinto River drainage instead of those for the Santa Ana River Drainage.

Table D2. Concentrations of constituents used to classify oxidation-reduction conditions in groundwater, Upper Santa Ana Watershed study unit, California GAMA Priority Basin Project.

[anoxic/suboxic, dissolved oxygen < 0.5 mg/L; indeterminate, insufficient data to determine redox classification; mg/L, milligrams per liter; µg/L, micrograms per liter; oxic, dissolved oxygen greater than or equal to 0.5 mg/L; redox, oxidation-reduction; <, less than; E, estimated; CDPH, California Department of Public Health; O_2, oxygen; Fe(III), iron; SO_4, sulfate; NO_3, nitrate; >, greater than; Mn (IV), manganese]

USGS GAMA well identification number	CDPH GAMA well identification number[1]	pH	Dissolved oxygen (mg/L)	Oxidizing and reducing constituents				Redox category	Redox process
				Nitrate plus nitrite (mg/L)	Manganese (µg/L)	Iron (µg/L)	Sulfate (mg/L)		
USAWB-01	USAWB-DG-01	7.8	1.2	8.4	<20	<100	63.0	Oxic	O_2
USAWB-02		7.3	8.0	1.1	<0.2	<6	21.3	Oxic	O_2
USAWB-03	USAWB-DG-03	8.1	4.7	2.1	<20	<100	100	Oxic	O_2
USAWB-04		7.9	10.0	2.7	<0.2	<6	16.9	Oxic	O_2
USAWB-05	USAWB-DG-05	7.5	9.8	1.8	<20	<100	31.0	Oxic	O_2
USAWB-06	USAWB-DG-06	7.8	8.1	5.0	<20	<100	35.0	Oxic	O_2
USAWB-07		7.7	5.5	4.9	<0.2	<6	74.2	Oxic	O_2
USAWB-08		7.4	8.8	10.5	<0.2	<6	53.0	Oxic	O_2
USAWB-09	USAWB-DG-09	7.2	20.2	3.2				Oxic	O_2
USAWB-10		7.6	9.6	5.5	0.3	<6	61.7	Oxic	O_2
USAWB-11		7.3	8.1	7.5	0.3	5.0	116	Oxic	O_2
USAWB-12		7.3						Unknown	Unknown
USAWB-13	USAWB-DG-13	7.5	11.0	15	<20	<100	62.0	Oxic[2]	O_2
USAWB-14		7.4		2.3	<0.2	<6	17.2	Oxic	O_2
USAWB-15	USAWB-DG-15	7.5	4.0	4.5	<20	120.0	28.0	Oxic or mixed[3]	Fe(III)/SO_4
USAWB-16		7.8	7.9	1.6	0.2	4.0	38.2	Oxic	O_2
USAWB-17	USAWB-DG-17	7.5	11.1	3.4	<20	<100	20.0	Oxic	O_2
USAWB-18		7.7	7.7	6.0	E0.1	<6	30.5	Oxic	O_2
USAWB-19	USAWB-DG-19	7.5	7.8	7.0	<20	170	29.0	Oxic or mixed[4]	Fe(III)/SO_4
USAWC-01		8	7.8	5.4	<0.2	<6	16.6	Oxic	O_2
USAWC-02		8	7.7	3.0	E0.1	<6	8.4	Oxic	O_2
USAWC-03	USAWC-DG-03	7.6	9.1	2.9	<0.2	<20	18.0	Oxic	O_2
USAWC-04		7.5	9.1	15	E0.1	<6	85.4	Oxic	O_2
USAWC-05	USAWC-DG-05	7.5	10.6	16	<0.2	<20	26.0	Oxic	O_2
USAWC-06	USAWC-DG-06	7.4	7.8	2.5	0.3	<20	38.0	Oxic	O_2
USAWC-07	USAWC-DG-07	7.6	0.7	0.8	<20	<100	120	Oxic	O_2
USAWC-08		7.7	8.1	12	<0.2	6.0	26.8	Oxic	O_2
USAWC-09	USAWC-DG-09	8.6	0.2	0.0	22	170	38.0	Anoxic	Fe(III)/SO_4
USAWC-10		8.2	10.6					Oxic	O_2
USAWC-11		7.4	6.2	5.4	2.0	6.0	60.4	Oxic	O_2
USAWC-12		7.5	4.7	34.8	10	6.0	42.2	Oxic	O_2
USAWC-13	USAWC-DG-13	7.6	6.5	1.7	45	380	36.0	Oxic or mixed[4]	Fe(III)/SO_4
USAWC-14		7.5	7.7	6.4	E0.1	<6	18.7	Oxic	O_2
USAWC-15	USAWC-DG-15	7.8	5.8	3.8	<20	<100	52.0	Oxic	O_2
USAWC-16	USAWC-DG-16	7.9	8.2	3.2	<20	<100	27.0	Oxic	O_2
USAWC-17		7.7	9.6	2.4	<0.2	<6	13.7	Oxic	O_2

Table D2. Concentrations of constituents used to classify oxidation-reduction conditions in groundwater, Upper Santa Ana Watershed study unit, California GAMA Priority Basin Project.—Continued

[anoxic/suboxic, dissolved oxygen < 0.5 mg/L; indeterminate, insufficient data to determine redox classification; mg/L, milligrams per liter; µg/L, micrograms per liter; oxic, dissolved oxygen greater than or equal to 0.5 mg/L; redox, oxidation-reduction; <, less than; E, estimated; CDPH, California Department of Public Health; O_2, oxygen; Fe(III), iron; SO_4, sulfate; NO_3, nitrate; >, greater than; Mn (IV), manganese]

USGS GAMA well identification number	CDPH GAMA well identification number[1]	pH	Oxidizing and reducing constituents					Redox category	Redox process
			Dissolved oxygen (mg/L)	Nitrate plus nitrite (mg/L)	Manganese (µg/L)	Iron (µg/L)	Sulfate (mg/L)		
USAWC-18	USAWC-DG-18	7.9	7.9	12.7	<20	160	42.0	Oxic or mixed[4]	Fe(III)/SO_4
USAWC-19	USAWC-DG-19	7.7	7.0	0.8	<20	<100	13.0	Oxic	O_2
USAWC-20		7.8	7.8	4.1	<0.2	<6	9.5	Oxic	O_2
USAWC-21		7.5	10.1	5.9	0.5	3.0	41.5	Oxic	O_2
USAWC-22		7.6	7.6					Oxic	O_2
USAWC-23		7.7	8.8	8.4	<0.2	3.0	34.6	Oxic	O_2
USAWC-24		7.9	2.2	1.3	2.4	8.0	17.9	Oxic	O_2
USAWC-25		7.4	5.7	13.9	E0.1	<6	67.6	Oxic	O_2
USAWE-01		8	5.6	1.4	0.4	10.0	77.5	Oxic	O_2
USAWE-02		9.2	0.3	0.1	3.8	<6	81.4	Anoxic	Suboxic
USAWE-03		7.4	1.2	4.5	28	<6	167	Oxic	O_2
USAWE-04		7.1	8.8	4.3	0.5	<6	136	Oxic	O_2
USAWR-01	USAWR-DG-01	7.4	7.4	11.3	<20	<100	42.0	Oxic	O_2
USAWR-02	USAWR-DG-02	8	6.9	5.4	<20	<100	30.0	Oxic	O_2
USAWR-03		7.3	6.1	12.9	<0.2	<6	97.8	Oxic	O_2
USAWR-04	USAWR-DPH-7	8.5	6.6	5.9	<20	<100	76.0	Oxic	O_2
USAWR-05		7.1	2.0	10.4	0.6	<6	80.7	Oxic	O_2
USAWR-06		7.4	5.8	19.0	0.2	18.0	180	Oxic	O_2
USAWR-07		7.2	3.6	17.9	<0.2	<6	232	Oxic	O_2
USAWR-08		7.1	0.2	9.2	E0.1	3.0	101	Anoxic	NO_3
USAWR-09		7.2	6.6	16.2	<0.2	<6	78.2	Oxic	O_2
USAWR-10		7.2	2.1	3.9	0.2	3.0	189	Oxic	O_2
USAWR-11		7.8	10.8	5.6	0.2	6.0	23.7	Oxic	O_2
USAWR-12		7.3	7.9	17.5	0.8	11.0	164	Oxic	O_2
USAWS-01	USAWS-DG-01	7.1	7.9	10.2	<20	<100	36.0	Oxic	O_2
USAWS-02		7.7	6.4	11.0	0.4	10.0	41.2	Oxic	O_2
USAWS-03		7.6	6.1					Oxic	O_2
USAWS-04		7	3.8					Oxic	O_2
USAWS-05		6.9	9.3	6.6	1.7	34.0	196.0	Oxic	O_2
USAWS-06		7.4	0.2	0.1	275	441	3.9	Anoxic	Mn(IV) and Fe(III)/SO_4
USAWS-07		6.6	6.3					Oxic	O_2
USAWS-08		6.8	9.6	13.7	1.1	10.0	22.9	Oxic	O_2
USAWS-09	USAWS-DG-09	7.2	7.5	3.8	<20	<100	55.0	Oxic	O_2
USAWS-10		7.7	7.5					Oxic	O_2
USAWS-11		7.2	6.6					Oxic	O_2
USAWS-12		7.1	10.5	5.4	0.2	4.0	44.2	Oxic	O_2

Table D2. Concentrations of constituents used to classify oxidation-reduction conditions in groundwater, Upper Santa Ana Watershed study unit, California GAMA Priority Basin Project.—Continued

[anoxic/suboxic, dissolved oxygen < 0.5 mg/L; indeterminate, insufficient data to determine redox classification; mg/L, milligrams per liter; µg/L, micrograms per liter; oxic, dissolved oxygen greater than or equal to 0.5 mg/L; redox, oxidation-reduction; <, less than; E, estimated; CDPH, California Department of Public Health; O_2, oxygen; Fe(III), iron; SO_4, sulfate; NO_3, nitrate; >, greater than; Mn (IV), manganese]

USGS GAMA well identification number	CDPH GAMA well identification number[1]	Oxidizing and reducing constituents						Redox category	Redox process
		pH	Dissolved oxygen (mg/L)	Nitrate plus nitrite (mg/L)	Manganese (µg/L)	Iron (µg/L)	Sulfate (mg/L)		
USAWS-13		7	7.0		32	5.0	43.1	Oxic	O_2
USAWS-14		8.1	0.2	0.2	1.5	31.0	188.0	Anoxic	Suboxic
USAWS-15		7.5	0.8	3.4	1.6	5.0	152.0	Oxic	O_2
USAWS-16		8.2	0.6	0.6				Oxic	O_2
USAWS-17		7.9	3.2					Oxic	O_2
USAWS-18		7.5	2.5	14.6	5.2	26.0	260.0	Oxic	O_2
USAWS-19		9.3	0.2				119.0	Anoxic	Unknown
USAWS-20		7.5	4.7	7.0	0.4	8.0	43.8	Oxic	O_2
USAWS-21		8.7	0.2	<0.1	115	82.0	20.5	Anoxic	Mn(IV)
USAWY-01	USAWY-DG-01	7.6	8.5	3.5	<20	<100	20.0	Oxic	O_2
USAWY-02	USAWY-DG-02	7.8	6.7	3.4	<20	<100	41.0	Oxic	O_2
USAWY-03	USAWY-DG-03	7.5			<20	<100	33.4	Oxic[5]	O_2
USAWY-04		7.7		1.5	<0.2	<6	51.0	Oxic[2]	O_2
USAWY-05		7.7	6.2	5.6	0.2	5.0	20.9	Oxic	O_2
USAWY-06		7.7	8.8	7.4	E0.1	<6	30.0	Oxic	O_2
USAWY-07	USAWY-DG-07	7.2	6.5	5.4	<20	<100	36.0	Oxic	O_2
USAWY-08	USAWY-DG-08	7.6	4.8	2.0	<20	<100	29.0	Oxic	O_2
USAWY-09	USAWY-DG-09	7.8	10.2	0.7	<20	<100	23.0	Oxic	O_2
None	USAWB-DPH-17	7.1		7.5	<20	<100	130.0	Oxic[2]	O_2
None	USAWC-DPH-1	7.8		12	<20	<100	69.0	Oxic[2]	O_2
None	USAWC-DPH-10	7.3		90	<20	<100	76.0	Oxic[2]	O_2
None	USAWR-DPH-7	8.5		5.9	<20	<100		Oxic[2]	O_2
None	USAWY-DPH-1			<0.5				Unknown	Unknown
USAWU-01		7.4	0.8	4.9	42	<6	79.2	Oxic	O_2
USAWU-02		7.6	4.8	4.4	<0.2	<6	33.1	Oxic	O_2
USAWU-04		7.7	7.5	7.2	<0.2	8.0	32.8	Oxic	O_2
USAWU-05		7.5	8.5	8.6	0.2	8.0	52.8	Oxic	O_2
USAWU-06		7.5	9.4	8.6	<0.2	<6	38.3	Oxic	O_2
USAWU-07		7.6	7.9	7.1	<0.2	<6	20.7	Oxic	O_2
USAWU-08		9.2	0.3	1.2	1.9	7.0	25.1	Anoxic	NO_3
USAWU-09		7.2	6.7	21.8	<0.2	<6	44.2	Oxic	O_2

[1] Samples with a CDPH GAMA identification number are those for which data were obtained by CDPH.

[2] Sample has no DO data, and is presumed to be oxic. It could be anoxic (NO_3-reducing).

[3] Sample has no DO data and CDPH Fe > 100.

[4] Sample has DO > 0.5 (Oxic), but CDPH Fe data may indicate Fe-reducing.

[5] Sample has no DO data and is presumed to be oxic. It could be anoxic (suboxic or NO_3-reducing).

www.ingramcontent.com/pod-product-compliance
Lightning Source LLC
Chambersburg PA
CBHW081551170526

45166CB00009B/2655